EXPERIENCING NATURE

Experiencing Nature

THE SPANISH AMERICAN EMPIRE AND THE EARLY SCIENTIFIC REVOLUTION

ANTONIO BARRERA-OSORIO

University of Texas Press *Austin*

Printed in the United States of America
First edition, 2006

Requests for permission to reproduce material from this work should be sent to:
Permissions
University of Texas Press
P.O. Box 7819
Austin, TX 78713-7819
www.utexas.edu/utpress/about/bpermission.html

♾ The paper used in this book meets the minimum requirements of ANSI/NISO Z39.48-1992 (R1997) (Permanence of Paper).

Barrera-Osorio, Antonio
Experiencing nature : the Spanish American empire and the early scientific revolution / Antonio Barrera-Osorio. — 1st ed.
p. cm.
Includes bibliographical references and index.
ISBN 0-292-70981-1 (cl. : alk. paper)
1. Science—Latin America—History—To 1830. 2. Science—Spain—History—To 1830. 3. Science—United States—History—To 1830. 4. Latin America—History—To 1830. 5. Spain—History—To 1830. 6. Spain—Colonies—America. I. Title.

Q127.L29B37 2006
509.8—dc22

2005011166

A mi mamá, Rocío, quien me dio todas las oportunidades del mundo;

a mi papá, Antonio, quien murió hace ya muchos años pero a quien todavía echo de menos;

a Rodrigo y Marta, que ya sabían que este libro sería;

y a Pilar, por supuesto.

For my mother, Rocío, who made it all possible;

for my father, Antonio, who died many years ago and whom I still miss very much;

for Rodrigo and Marta, who knew that this book was going to happen;

and for Pilar, of course.

Contents

Acknowledgments

This book grew out of my Ph.D. dissertation at the University of California at Davis under the direction of Paula Findlen, who has become a good friend over the years. I thank her for her trust, fine teachings, sharp criticisms, mentorship, and friendship. Pamela Smith became closely involved with the project, always asking questions that forced me to think beyond Spain. Chuck Walker and Arnie Bauer were also part of my dissertation committee and, as such, part of this book. From them I learned about Colonial Latin America, archives, and the joys of learning about other cultures. I had two dear professors, Tom Brady and Betty Jo Dobbs, who taught me their love for history with a sense of humor and human compassion. I thank all of them for their comments and support over the years. Many other friends and colleagues will recognize their marks, and I thank them again here. At the dissertation stage I received support from the Fulbright Commission as well as the University of California at Davis. I also received financial help from the Research Council of Colgate University. I thank these institutions as well.

The book took me from Seville to Providence, where I began and ended the research for it. I went to Seville to pursue a project on natural history and came back with a completely different project on empirical practices. I went to Providence to research a new project, which I did, but ended up collecting new information (especially the illustrations) for this book as well. Seville and Providence have one element in common, apart from people like me: they belong to the Atlantic world. Both Seville and Providence hold stories about each other and about the larger world of the

Atlantic. Both house great archives, the Archivo General de Indias (AGI) and the John Carter Brown Library (JCB), each distinctive yet so similar in their Atlanticism, scope, and unpredictability.

I thank the archivists and librarians at the AGI and the JCB: in Seville, Jesús Camargo, Magdalena Canilla, Isabel Ceballos, María Antonia Colomar, Belén García, Pedro González, Rafaela González, Pilar Lázaro de la Escosura, Purificación Medina, Blanca Yrazusta, Antonio López, María Teresa Molina, and Socorro Prous. Socorro not only helped me understand the archive; she and her family adopted my family and supported my research, and us, in countless ways. In Providence, I thank the director Norman Fiering as well as the librarians, especially Susan Danforth, Denis Landis, Richard Ring, and Valeria Gauz, for their suggestions and help with my projects. Nan Summer-Mack, Adelina Axelrod, Valerie Andrews, Jennifer Gage, Abigail Harkey, Lynne Harrell, Richard Hurley, Heather Jespersen, Valerie Andrews, Valerie Krasko, and Susan Newbury made me feel welcome and at home in Providence. Thanks for your good work and good sense of reality. I owe thanks as well to the archivists and librarians at the Archivo Nacional de Madrid, the Biblioteca Nacional de Madrid, the Biblioteca Colombina in Seville, the Biblioteca de la Universidad de Sevilla, the Bancroft Library in Berkeley, and the Library of Congress in Washington.

Talks based on the book were presented at the History of Science Society in Denver, Milwaukee, and Boston as well as in the Colloquium in the History of Science at the University of California, Berkeley. I am grateful to the members of these audiences for their comments and questions. I am also grateful for the encouragement and support of my colleagues in the History Department at Colgate University. Two copyeditors, Lauren Osborne and Rebecca Costello, helped me with the manuscript: Lauren in the early phases of the manuscript project and Rebecca in the final stage. I thank them for their wonderful and well-crafted work. Kathy Lewis did an excellent job with the final manuscript for the University of Texas Press in both English and Spanish—*muchas gracias.* I want to acknowledge as well the careful reports of the readers for the University of Texas Press; and I wish to express a personal note of appreciation to Theresa May and Allison Faust, who supported this project from the very beginning. After so many years of research, readings, conversations, and talks, this project is now a book, thanks to your trust and support: I am profoundly grateful to you.

Finally, thanks to my family and friends. I have been dreaming about this thank-you letter to them for a while. I am pleased it is part of this book.

I thank my mother, Rocío Osorio, and my brothers, Andrés and Felipe, for their love and good words and for listening to these alien stories with intelligence and humor. Rodrigo and Marta Mejía, my parents-in-law, did not live to see this book, but they have been part of it from the beginning. Rodrigo was a perfect interlocutor, with his questions and profound skepticism. He pushed me to think clearly about issues and problems. Marta was always there for us. I dedicated my dissertation to them. That part of this book is still for them. I especially want to thank Pilar, who has always been there helping me to think better and who has done so much so that I can do what I am doing. I am eternally grateful to you: *Gracias, socia, con todo mi amor.* To my children, Antonio and Candelaria, I owe a thousand and one thanks for their love and playing time, for their smiles, for their questions, and for listening to my stories. I am trying to connect the past with the future for them, without exactly knowing how.

Gracias to all of you: teachers and colleagues, archivists and librarians, editors, friends, and family—*muchas gracias.*

Introduction

In 1592 King Philip II of Spain received a sample of a tree with aromatic properties from Puerto Rico.[1] The governor of Puerto Rico had sent the sample, suggesting that physicians study those properties to determine the tree's economic value. Philip II ordered his physician, Doctor Mercado, to conduct tests and report back to the president of the Council of Indies. On the basis of this report, the president of the council would make a decision about the economic potential of the tree and then inform the governor of Puerto Rico.

In this book, I explore the history of scientific experience during the sixteenth century and examine various aspects of its development through the study of the Spanish empire. By the late sixteenth century, royal institutions and a network of bureaucrats, royal officials, physicians, merchants, adventurers, pilots, and friars from Spain were circulating information about natural entities and samples from the New World. Through these activities, a set of rules and practices for the collection, organization, and dissemination of information regarding the natural world of the Indies developed. I look at the institutions established for conducting this work; the mechanisms adopted for testing and establishing accurate information; and the ways in which empirical information was used for the production of new knowledge. From a larger perspective, I seek to integrate the Atlantic world into the history of science.

The Atlantic world fostered the development of one key element of modern epistemological practices: empirical observation.[2] Numerous accounts and descriptions of the New World, together with the increased circulation of natural entities such as my opening example, helped to establish this empirical tradition, which, in its turn, helped to break the late medieval and humanist dependence of knowledge upon textual interpretation and exegesis.[3] The establishment of this tradition took place during a crisis of authority and the rise of material culture in Europe. In both events, the Atlantic world, and in particular Spanish America, played a significant and decisive role.

This was the Spanish-American contribution to the development of sixteenth-century science: the institutionalization of empirical practices at the House of Trade and Council of Indies, together with the books that were then written about these practices. These books would eventually arrive in England, where natural historian Francis Bacon would continue this tradition in the next century and launch an empirical program at the Royal Society of London tantalizingly similar to the one launched in Spain. The Spanish contribution to the development of science consisted of the institutionalization of empirical practices rather than the theoretical development of science—those developments would come only during the seventeenth century in England, Holland, and France and by the late seventeenth century in Spain.

I refer to the emergence of empirical practices and their institutionalization as the "early Scientific Revolution." The Scientific Revolution did not start with Nicolaus Copernicus and his heliocentric ideas, or with the publication of books by artisans and painters. I argue that it started in the 1520s, in Spain, when merchants, artisans, and royal officials confronted new entities coming from the New World and had to devise their own methods to collect information about those lands: there were no avocados in Pliny's pages.

Many events fostered the aforementioned crisis of authority taking place during this period, which made conditions ready for this early Scientific Revolution. Humanists unearthed books and texts that, when printed, opened up the horizon of cultural cognitive models. Popular and noble revolts from Germany to Spain questioned royal authority and proposed new ways of exercising political power, bringing new social groups with their own claims about social order into the courts and city councils. Mystics, monks, and saints everywhere, from the Spanish Iluminados to Martin Luther, tried—or managed—to reform religious practices and redefine the role of God in human and natural affairs. Protestants and Catholics brought new ideas about the Christian God and new religious practices—and invented new forms of persecution. National and religious identities not only reshaped international relations but also confessionalized religious practices. Within this context of fragmentation, Atlantic travels brought news from hitherto unknown lands. People began to question the great classical narrative about the natural world.[4]

It is well known that European naturalists during this period developed different programs for investigating discoveries in the New World. Following Aristotle, Pliny, and the classical medical tradition, some naturalists emphasized empirical evidence and the collection of new information and items to confirm and support their texts.[5] Others emphasized Aristotle's conception of essences and developed a method to determine the essences of natural entities, as the physician and botanist Andrea Cesalpino (1519–1603) did with plants. The Hermetic tradition, dealing with astrology and magic, had many followers. The physician and alchemist Paracelsus (1493–1541) and his disciples rejected traditional medical authorities to rely only on a Christian approach to nature based on empirical observations.[6] In connection with the commercial culture of the sixteenth century, alchemists developed their own program of research based on the empirical study of natural products to control hidden natural forces.[7] Sixteenth-century scientific practitioners began to leave aside traditional textual-medieval practices and to search for empirical methods of understanding nature.

Perplexingly, historians of science tend to overlook the program for researching nature developed by the Spaniards in the Atlantic world. Yet during the sixteenth century Spain mobilized most of its energies in search of riches and domination.[8] A by-product of this mobilization was the emergence and institutionalization of empirical practices; thus, the story of how these practices emerged is interconnected with the politics and economics of the country in that period.

To establish the setting in which the events of this book occur, I provide a brief description of the political events of the period. In 1516 Charles of Habsburg (1500–1558) was proclaimed king of Castile and the Spanish kingdoms. His dominions extended to the Netherlands, the Spanish possessions in Italy, and the New World. In 1519 he was elected Holy Roman emperor. The Castilian crown came to him because his mother, Juana of Castile, became mentally unstable after the death of her husband, Philip I, who ruled Spain for a year (1506). The imperial crown came to Charles because he was able to obtain enough money from German bankers to buy the electors. Madness and money made possible Charles V's empire. But the Spanish kingdoms that Charles inherited came to be *those* kingdoms by design: Charles's grandparents, Ferdinand (1452–1516) and Isabella (1451–1504), had been working since the late fifteenth century on the unification and expansion of Spain. By 1492 the Catholic kings had expelled the Jews from Spain, captured Granada from the Moors, and sent Christopher Columbus on his voyage to the East.

After 1492 the Catholic kings began the expansion of royal power in Spain, in Italy, and in the New World. They were able to control Spanish town councils and religious orders by appointing their own representatives in those political and military bodies. The military orders, in particular, provided the crown with access to land and patronage. The appointment of royal officials to town councils extended the royal authority to cities, a source of revenue as well. By the time Isabella died (1504), the Spanish kingdoms were in the process of unification. Her son and heir, Juan, had died in 1497; and her grandchild Miguel, who could have united Spain and Portugal, had died in 1500. The heiress to the throne was now Juana (1479–1555), married to Philip of Habsburg. Ferdinand assumed the regency of Castile until 1506, when Philip I arrived in Spain. Philip died that year, "after drinking a glass of very cold water, sweating, at the end of a ball game."[9] Juana's mental instability gave Ferdinand the opportunity to control the crown as regent. When Ferdinand died in 1516, the crown passed to his grandson, Charles I (1500–1558), son of Juana and Philip.

Charles received several Spanish kingdoms, each with its own set of laws and traditions. His base was the kingdom of Castile, which provided most of the funds he needed for his imperial designs. Charles moved from a world where the preservation of empire and church had been an ideal worth fighting for into a world that was breaking them apart. He sought to preserve both the Catholic Church and the empire against the religious and state elements that were threatening them: Protestantism and central (expansionist) states. Charles fought against the Ottomans, who were expanding in central Europe and the Mediterranean; against Francis I, who sought to contain Charles in Italy; and against the Protestant lords in the Holy Roman Empire, who sought to establish their own areas of jurisdiction against the emperor. Charles used money from Castile, the Netherlands, and the New World to pay for his policies and wars.

Charles sought to increase his revenues by expanding commerce with the New World. By the early sixteenth century, entrepreneurs were looking for sources of revenue in the New World and were approaching the crown with proposals such as the commercialization of new medicines and the making of instruments (fishing rakes, for instance) for a more efficient exploitation of resources. From the interaction between entrepreneurs and royal officials emerged a set of empirical procedures to determine the value of these projects. Eventually these procedures were formalized in the royal institutions responsible for the government of the American kingdoms. Charles V's need for revenues and the entrepreneurs' willingness to find sources of revenue in the New World created the context in which empirical practices in the study of nature could emerge. Charles V's successor, Philip II (1556–1598), inherited his goals, his wars, and his constant need for resources to build his European world order.

Philip II was thus constantly pressed for funds: from the 1560s to the 1570s Spain was engaged in wars in France, the Mediterranean, and the Low Countries—and this war would last beyond his reign. In the 1580s Spain annexed Portugal and sent an armada against England; Philip needed funds to accomplish this. The New World was a source of wealth (primarily silver but also potential commodities), and Philip continued his father's policy of fostering and supporting commercial activities there. The center of these activities was the search for commodities and the improvement of previous technologies and instruments. These activities were based in turn on the empirical study of nature. Philip II's reign saw the consolidation of empirical scientific activities that emerged in the context of commercial expansion and empire formation under Charles V. The literature on the Scientific Revolution does not take into account the significance of this consolidation. On the contrary, the silence about Spain is perplexing: it was the ruling kingdom in Europe during the sixteenth century—and it did make a lot of noise.[10]

The Italians, French, and English were constantly interested in Spain's activities in the New World, especially its scientific pursuits. For instance, books by Pedro de Medina and Martín Cortés were translated into Dutch, French, Italian, and English; in 1558 Stephen Borough visited the navigational school of the Casa de la Contratación in Seville with the intention of establishing a similar institution in England; and the natural historian Carolus Clusius traveled in Spain in the 1560s, collecting samples and books and establishing professional relations with Spanish natural historians. Spanish books were translated into other European languages; foreigners visited Spanish institutions; and Spanish scholars received visits and letters from foreign scholars. Sixteenth-century Spain was at the forefront of the development and institutionalization of empirical activities in Europe.[11]

Spanish efforts to control the New World were also part of an effort to control the fragmentation in authority taking place in Europe. By establishing an empire in the New World and opening new sources of revenue there, the Spaniards sought to maintain their predominance in Europe—fighting Protestants, protecting Europe from the Ottomans, and fostering

Catholic religious practices everywhere. There was a strong sense of mission in the Spanish enterprise of the sixteenth century, linking Tenochtitlán with Vienna via Seville—all centers of Charles V's empire.

Thus, with the establishment of the Spanish empire in the New World came the need for a practical understanding of the natural world there through institutions, practices, and mechanisms for exploring nature, mapping new lands and the oceans, and collecting commodities, curiosities, and information.

Royal authorities and merchants did not know anything about the New World—about its geography, natural history, or peoples. They needed practical information. From the point of view of the Spanish rulers and people engaged in the American enterprise, the accumulation of empirical evidence constituted a sensible basis for political and economic decisions and for dealing with the increased flow of things and information circulating in the Atlantic world.[12] Spaniards began to study the New World: using, collecting, and organizing empirical information and collaborative practices and institutionalizing them at the Council of Indies and the House of Trade.

The development of empirical practices was not only the result of long-distance strategies for controlling the New World. At the time of its early encounters there, Spain already had rich cultural and intellectual resources. Spanish courts and universities (such as Salamanca and Alcalá) discussed humanist and nominalist ideas, exposing the clergy and the royal bureaucrats who traveled to the Indies in the early years to these ideas. For example, the humanists Gonzalo Fernández de Oviedo (1478–1557) and Bernardino de Sahagún (ca. 1499–1590), who collected empirical information about the New World and wrote natural histories, both came out of this cultural milieu.

In addition, sixteenth-century Spain had inherited rich Arabic scientific traditions. After seven centuries of Arabic domination (700s to 1400s), Spanish scientific culture had been strongly affected by Arabic science, particularly in astronomy and medicine. The astronomical tables known as the Alfonsine Tables, which were part of a corpus based on medieval Arabic astronomy, constituted one of the most important astronomical almanacs in Europe until the mid-sixteenth century. Arabic books on medicine and philosophy were preserved at the University of Alcalá, according to the instructions of its founder, Cardinal Francisco Jiménez de Cisneros, who ordered the burning of all other Arabic texts after the fall of Grenada. This inheritance was reinforced by Jewish investigations in medicine and cosmography. Both influences were still present in sixteenth-century Spain. By the time Spain began to colonize America, it had at its disposal cultural resources favorable for the development of natural history, cosmography, navigation, medicine, and mathematics.

These scientific activities became indispensable in mastering the ocean and lands of the Atlantic world. Yet, in that context, practices were reinvented in light of information coming from there and its practical uses rather than in relation to classical texts. In this new configuration of aims and interests, new mechanisms for gathering, organizing, and disseminating information about the world were developed within activities such as natural history and cosmography. Classical traditions became relative points of reference in an ever-increasing circuit of information.

Most of the scholarship on changing conceptions of nature in the wake of the Columbian accident has focused exclusively on published natural histories: in this book, I contextualize this body of literature in its broader setting by exploring unpublished materials. The idea of writing the natural history of the Indies, for instance, grew out of a collective enterprise: Gonzalo Fernández de Oviedo's and José de Acosta's histories were actually complex pastiches, involving compilations, glosses, and translations of a number of different sources.

I recount two overlapping stories. The first is the story of how empirical practices emerged from the relationship between the crown and its subjects. Imperial and commercial activities such as navigation, agriculture in the New World, and instrument-making fostered a culture based on experiential and collaborative practices supported by royal officials and merchants. These activities—conducted by people representing a variety of causes and concerns—intersected with each other and included the promotion of practices such as sending questionnaires and expeditions to gather empirical information, requesting reports from witnesses, establishing juntas of experts for organizing information, and performing tests to determine claims of truth. Out of the convergence of these imperial activities around the New World emerged this empirical culture that, in turn, supported the development of modern science. And while the crown established institutions to maintain and promote these cultural practices, these practices did not depend solely on state institutions—merchants, explorers, and royal officials also participated in their creation and development.

The second story details how the crown institutionalized these empirical practices and their embodiment in procedures and methods. These practices and institutions were linked to the development of Spain's long-distance strategies for controlling the New World. At first they were an unintended consequence of interconnected actions, emerging from state officials' and private actors' plans, decisions, and projects. But later the crown appropriated many private initiatives and launched them as state projects, beginning the process of institutionalizing mechanisms and offices for gathering, organizing, and disseminating information. In fact, empirical practices eventually constituted a strategy for controlling, exploring, exploiting, collecting, and studying the nature of the New World—a dimension that is still part of science today.

My work builds upon the work of the Spanish historians José María López Piñeros and Raquel Alvarez and the work of David Goodman, among others.[13] These authors, building upon José Antonio Maravall, stress the connection between the Spanish crown and the development of scientific fields in Spain. López Piñeros is the leading figure in the study of the history of science in Spain during the sixteenth and seventeenth centuries; Raquel Alvarez studies the development of science in the New World; and David Goodman looks at the scientific practices supported by Philip II. My aim is to understand how empirical practices became the tool to study nature—how personal experience became the center of knowledge production. Contrary to Robert Merton, who "advanced the hypothesis that Puritanism in particular and ascetic Protestantism in general served to legitimate the new science,"[14] I argue that the commercial and empire-building culture of this period legitimized the new empirical practices of the new science. Modern science was the result of state and commercial activities, which did legitimize the new practices.

Edgar Zilsel argues that the decisive factor in the emergence of modern science was the interaction between artisans and scholars, through which scholars began to hold artisans and their methods in higher esteem than in the past.[15] I argue that this process took place within the European imperial and commercial expansion of the sixteenth and seventeenth centuries. Merchants, royal officials, artisans, natural historians, pilots, and cosmographers came together in institutions such as courts and academies, where their economic and political interests overlapped.

The Casa de la Contratación of Seville is a clear example: in the early sixteenth century it became the place—the institution—for pilots, cosmographers, and experts to produce knowledge about the New World. The Casa established a veritable Chamber of Knowledge, the first of its kind—in many ways different from the ancient model of Solomon's House as a place of knowledge.[16] For example, pilots and cosmographers hired by the Casa needed to interview pilots and travelers to the New World and used the information they collected for the making of charts. In many cases, the

Casa pilots and cosmographers had to organize special meetings (juntas) with the regular pilots to determine locations and routes for the making of charts. In both cases, through interviews or in meetings, it was in the Casa that these actors produced knowledge about the New World. This institution emerged as the result of Spain's commercial and imperial expansion in the New World—and there was not anything Puritan about it.

My work also expands upon the work of Lucile Brockway and Mary Louis Pratt, who argue that scientific practices in the eighteenth and nineteenth centuries served economic and political goals. Brockway studies the case of botanical practices in London and the English Asian colonies; and Pratt studies European expeditions and sample-collecting practices in Spanish America and how these practices helped to foster imperial and commercial interests. I place my argument before the institutionalization of modern scientific practices: it considers problems such as the validation of empirical practices and how empirical practices, in particular, emerged within the context of commercial and imperial expansion.[17] Once these practices were institutionalized, they continued to serve the purpose of empires and commercial groups (and nations, for that matter: for instance, the newly created American nations in the eighteenth and nineteenth centuries), as Brockway and Pratt have shown.[18] I argue that empirical science was originally a product of commercial and imperial expansion—then, once it was institutionalized, science served, in turn, the interests of the empire and nation states.

The chapters of this book are thematic; the goal is to build a thick narrative about the emergence of empirical practices by adding a new layer to the previous discussion with each chapter. The emergence and institutionalization of empirical practices in Spain (and, for that matter, in Europe) did not concern one single aspect, such as navigation. Instead, the emergence of empirical practices became rooted in several different areas, such as navigation and natural history, instrument-making and patents, and questionnaires and reports. While repeatedly covering a similar chronology within the sixteenth century, each chapter adds another layer to the understanding of how empirical practices emerged during that period. Although the chapters can be read independently, it is only together that they sustain the larger claim presented in this book.

I focus on five areas of Spanish activity in America: the search for commodities in and the ecological transformation of the New World (Chapter 1); the institutionalization of navigational and information-gathering practices at the Casa de la Contratación (Chapter 2); the development of instruments and technologies for the exploitation of natural resources in the New World (Chapter 3); the use of reports and questionnaires for gathering information (Chapter 4); and the writing of natural histories about the Indies (Chapter 5). In general, the Spanish royal institutions responsible for the American enterprise adopted and enhanced the idea of empiricism; published and unpublished reports and chronicles of the New World sharpened this idea. Their practices were shaped largely by the needs of the market and the exploitation of natural resources for commerce, especially minerals, tropical commodities, and medicinal plants. My approach shows that the activities discussed in this book were not isolated practices but rather the result of changes transforming the way in which Spaniards in the New World studied and approached the natural world.

By the early seventeenth century the demands of empire surpassed the resources of Castile and the New World. Spanish scientific activities remained rooted in commercial and political interests without completely establishing its goals, rules, and epistemological conditions. By the mid-seventeenth century the political reality of the European and the Atlantic world had become too mobile and uncertain for the Spaniards to direct its course. The growing complexity of the Atlantic world produced a remarkable change in Spanish attitudes toward nature, knowledge, and empire. In the sixteenth century the Spanish felt that they could control the Atlantic. In the seventeenth century they realized that they could barely maintain control over a portion of it. The *arbitrista* (economist) Martín González de Collerigo defined early-seventeenth-century Spain in these terms: "It seems that these kingdoms had been reduced to a republic of enchanted men who live outside the natural order."[19]

The political and economic uncertainties of the period, exacerbated by new wars against France, England, and the Low Countries, resulted in declining support for projects that would have enjoyed support in the previous century, such as the steam machines invented by Jerónimo de Ayanz (1605). Rooted in international economic and political contexts, Spanish science enjoyed variable support. Moreover, the Spaniards of the seventeenth century had lost faith in their own ability to transform the world around them. Since Spain's science was distinctively about transforming and controlling nature, it, too, lost its sense of purpose in this changing climate.

A sense of American identity (e.g., Mexican or Peruvian identity), however, began informing the work of natural historians and physicians. From the late sixteenth century onward, texts dealing with natural knowledge highlighted American themes and problems. The physician Agustín

Farfán, for example, wrote a treatise on medicine for the "benefit of this kingdom [of New Spain] and its republics, and to help poor people who lack the service of physicians."[20] Farfán used Doctor Francisco Hernández's work on natural history—written for the benefit of natural historians in Europe—but instead wrote his book for the benefit of New Spain and the "poor people" who lacked access to physicians.[21] Farfán identified his audience not only in New Spain but also in the larger group of poor people—making his knowledge accessible to a nonspecialized audience. Perhaps knowledge was not for all people, but certainly it was for many people in New Spain. By the end of the seventeenth century Spanish America was moving in a different direction than the Old Country. The empirical activities or early Scientific Revolution discussed in this book would eventually support the development of "enlightened" ideas in America; in other words, there was an internal culture already in place for the development of the Enlightenment in Spanish America. This book shows the establishment of that culture during the sixteenth century.

In conclusion, this book provides an approach to the study of the history of science that takes into account the influence of the Atlantic world in the development of sixteenth-century empirical practices. In the early sixteenth century most of the empirical practices emerging in the European context were bound to the humanist and Aristotelian understanding of empiricism or craft traditions. Yet how exactly did empiricism become a tool to understand and study nature? I suggest that the expansion into the New World constituted one element that allows for the understanding of this process. There is a chronological coincidence between European expansion and the institutionalization of empirical practices in European kingdoms. For instance, Bacon's proposals for the study of nature coincided with the English expansion into the New World. Through the Spanish translations that began in the 1550s with Richard Eden and continued in the 1620s with Samuel Purchas, Bacon knew about this empirical culture established around the study of the New World.[22]

Our understanding of the history of science is incomplete if it is restricted to Protestant achievements—as it briefly was in the heyday of the Merton thesis—or closed to European practices that do not situate the emergence of empirical practices in its Atlantic context. This work integrates into the history of science not only the history of the Spanish American enterprise but also the history of state formation in the development of science. Science is not always a neutral activity in pursuit of the truth but rather a political activity aimed at controlling nature. Most scholars of the history of science in the sixteenth and seventeenth centuries still continue to focus on developments solely within Europe. They are producing some of the most interesting work in history today, yet most of this work still fails to take into account the emergence of empirical practices in the context of European expansion. This book seeks to bridge that gap.

A final comment on terms is required. The term "America" refers to the American continents. This usage was already in place in the sixteenth century and is still current in many American countries, with the exception of the United States, where "America" means only the United States. The term "science" refers to natural history in particular but also to medicine and cosmography. I have tried to avoid the use of this term in the body of my work, because sixteenth-century scientific practices were more flexible, interrelated, and diffuse than our understanding of scientific practices is today.

ONE

Searching the Land for Commodities

In 1530 a new kind of balsam made its way to Spain from the New World. The crown ordered the merchants interested in exploiting balsam to send samples of it to physicians and hospitals. They, in turn, would send reports to merchants and royal officials:

> Physicians and surgeons of any city, town, and place of our kingdoms and possessions, before they talk or publish inexactitudes about this balsam, should have unequivocal information [*cierta noticia*] about it; and when, by experience or by other method, they find out that it is harmful for wounds or any other illness, they should declare and reveal it to our magistrates.[1]

The balsam case and others like it provide a window into the way in which sixteenth- and seventeenth-century Europeans understood and commercialized new natural products.[2]

In general, sixteenth-century Europeans understood nature as a repository of commodities, readily available for commercialization (as were so many other commodities in this era of discoveries).[3] By no means was this solely a European phenomenon; China, India, the Ottoman Empire, and Africa were perhaps more important players in the configuration of the world's trading framework.[4] Still, the Portuguese and Spanish explorations of the fifteenth and sixteenth centuries placed Europe on a path that would shape it from that point on.

The newly emerging European states were constantly looking for resources outside Europe, and the New World offered many new economic and political possibilities. European crowns and merchants established efficient mechanisms for economic and political control of the new lands. One consequence of these mechanisms was the emergence of empirical practices for understanding nature, expanding humanist and Aristotelian practices into new directions. Yet these humanist and Aristotelian practices constituted examples of empirical practices profoundly bound to textual authorities. The political and commercial expansion of this period created a context for the implementation of empirical practices free from textual authorities. In the case of Spain, both merchants and the crown encouraged the development of such practices, for both sensed that the empirical study of nature offered them material benefits.[5] The crown institutionalized these empirical practices in two bodies, the Casa de la Contratación and the Council of Indies.

In 1503 the crown established the Casa de la Contratación or House of Trade (perhaps based on plans by the royal official Francisco Pinelo) for the organization of trading relations with the New World, including the organization of the fleets and collection of royal tolls and taxes. A few years later, Ferdinand established a navigational and mapping center within the Casa for the organization of navigational activities, so that commerce could be conducted on a solid navigational basis. In 1524 the crown formally established the Council of Indies for the administrative and judicial organization of the New World. As happened with the Casa, the council would eventually establish offices for the collection of information about the New World (in particular natural history information) so that the administration of the New World could be conducted on the basis of solid information.[6]

Royal authorities needed specific information about the New World in order to control it: they needed to know about its geography and natural history, its peoples and types of governments. Entrepreneurs, by contrast, needed to know how to exploit the New World—so they sought information about the properties of the land and the uses of natural products. Both crown officials and entrepreneurs fostered the circulation of information and commodities, in which personal experience and testing became the criteria for validating information.

During this period of exploration, European notions of nature and experience were displaced from their traditional classical definitions. Natural history notions from classical authorities such as Aristotle or Pliny did not account for the size of the earth, a new continent, life in the

Torrid Zone, manatees and *mechoacán* (a Mexican medicinal herb). Experience, of course, had always played a role in validating knowledge; but the knowledge gained in exploration and in contact with other cultures in the New World made experience a much more important player than the authority of classical sources alone.

A CASE STUDY: THE SANTO DOMINGO BALSAM

Balsam was a celebrated classical medicine. According to Dioscorides in *De Materia Medica* (the foremost classical source on medicinal simples), Judea and Egypt produced balsam, but only in very small quantities. The liquor of balsam was better than its fruit or wood. The virtue of the liquor "was very effective, for its very hot quality."[7] It was used to cure vision problems, to purge, and to provoke menstruation and childbirth. It also helped to heal wounds, to provoke urine, and to mitigate fatigue; it was a good antidote against poison. Balsam was almost an all-purpose medicine, and in high demand. In the early fifteenth century the traveler Pero Tafur reported that the production of balsam was very limited, encouraging the selling of fake balsam.[8]

The garden of Matarea was, according to legend, the place where the Virgin Mary had found water when she and her son escaped to Egypt. This water irrigated the enclosed garden where balsam grew. According to the scholar Marcos Jiménez de la Espada, even a well-informed humanist such as Peter Martyr d'Anghiera (1457–1526) did not know that the production of the Matarea balsam had come to a halt until he visited the garden in 1502 and found that the fountain of water was dry.[9] Not long after, in 1528, Antonio de Villasante (a resident of Santo Domingo) claimed that he had found a similar product in the New World, in Hispaniola. He sought and received the crown's help in exploiting it. There was already a demand for balsam in the international drug market, and Villasante's discovery came just in time to supply it. Villasante's discovery also opened possibilities of revenues for the crown—in dire need of funds since the early 1520s, when Charles V began organizing campaigns against Francis I in Navarra, Guipúzcoa, and Italy.[10]

The discovery of the Hispaniola balsam, however, resulted not only from commercial demand but also from an informal search for commodities and curiosities in the New World. As early as the 1520s there was an interest in natural curiosities from the New World. In 1525 the humanist and royal official Peter Martyr obtained a royal decree to order shipmasters to bring animals and plants back to Spain.[11] In the same year, Charles V asked Gonzalo Fernández de Oviedo y Valdés (1478–1557) to write a natural history of the New World, which was published in 1526. Just few years later, in 1528, Antonio de Villasante presented his report on the new balsam before the Council of Indies.[12]

We know little about Antonio de Villasante, who was already a resident of Santo Domingo by 1514.[13] He received thirty-five Tainos in encomienda (a labor grant).[14] Villasante married Catalina de Ayahibex, a *cacica* (chief) who had converted to Christianity. He became friend of the viceroy Don Diego Colón (1479/80–1526) and later obtained a license from him to exploit balsam and other drugs on the island. Villasante traveled to Spain to secure a monopoly for the exploitation of this and other drugs (for example, rhubarb root and myrrh) in the Caribbean.

Villasante obtained from the crown the right to exploit balsam on the condition that he present before the Council of Indies:

> A long and very complete report about the tree to obtain the already mentioned liquor, and what its shape is and where this tree is found and what method is used to obtain the liquor; and similar [information] about other drugs.[15]

A few months later Villasante complied, presenting a report on Santo Domingo balsam before the council.

Villasante told the council that his wife, Catalina de Ayahibex, who was "an Indian, *cacica*, and Christian" (and therefore trustworthy in the eyes of Europeans), and "her family members" had taught him about the properties and uses of diverse medicinal plants from the island, including balsam.[16] He declared that he would state everything he knew about those plants, so that the king or his officials could eventually find and exploit them. The issue of secrecy became irrelevant, for Villasante's purpose was to obtain a commercial monopoly over balsam. It was to his benefit to provide all the necessary information for its commercialization and sale.

After establishing the authority of his source and information, Villasante continued with a description of balsam. He explained that he knew from experience that in Hispaniola, near Santo Domingo, there was a tree called *bálsamo* in Spanish and *boni, guacunax,* or *canaguey* in the native language, depending on the province. The tree was three yardsticks tall and grew near rivers and in wet areas; it was about as thick as a human arm. The big ones were bushy-topped; the leaf was very green and shaped like a rhombus. (Villasante provided a schematic drawing of a leaf.) The bark smelled and looked like cinnamon and tasted good, although it was a little hot and sour. The tree produced a fruit like that of the pepper tree but thicker.[17]

Villasante based his report on the authority and knowledge of the indigenous people and his direct access to them. He concluded that "the indigenous people affirm that there are many other beneficial trees and drugs in the Indies," and he promised to send further reports.[18] The Spaniards, of course, realized that they needed the knowledge of the indigenous people to survive and move around in the New World. In 1570 Philip II ordered his royal physician Francisco Hernández to travel to the New World to interview indigenous people and collect information about medicinal plants.[19]

Villasante, however, relied not only on indigenous knowledge but also on commercial interests. His intended audience, the Spanish bureaucracy and his commercial partners, shaped his description of balsam. It was no accident that he used the word *bálsamo* to translate the indigenous names *boni, guacunax,* and *canaguey,* for *bálsamo* was already the name of a valuable commercial drug. The physician Nicolás Monardes, years later, commented that this liquor "received that name because it produces great effects and cures many illnesses," as had the classical, Old World balsam.[20] But to make his point even more explicit, Villasante compared the Santo Domingo balsam with pepper and cinnamon, two extremely valuable spices. The New World, after all, had been encountered by Europeans precisely because of their perennial search for spices and medicines. In the few short years between Columbus and Villasante, merchants and the crown had extended their search for commercially viable new drugs and spices to include even plants and animals "of any quality and name." The Spaniards also took spices and medicines for agricultural and commercial purposes to the New World. Thus, Don Francisco de Mendoza, son of the first Mexican viceroy, in 1558 signed two capitulations with the princess Doña Juana (approved by Philip II in 1559) to cultivate ginger, sandalwood, pepper, cinnamon, and clover.[21]

Villasante confirmed the commercial possibilities of the Santo Domingo balsam by performing tests with it—what he called *esperiencias.* After the description of the tree, Villasante described the method for preparing balsam. He cut the branches with a knife, took off the leaves and seeds with his hand, and shredded the pruned branches together with bark from the trunk. Then he pounded this mixture with rocks and chopped it into pieces with a knife. Villasante warmed the mixture in a clay pot with water. Once the mixture had been soaked for a while, he took it out of the clay pot and squeezed it to obtain liquor. Finally, he heated the liquid in a small pot, which was inside a bigger one full of ashes, until it was reduced to thick liquor—the Santo Domingo balsam.[22] Sometimes he would put it out to dry under the sun, with the same results. When he cut the trunk with a knife, a "liquor" came from the tree. This sap "as it was coming out, hardened like gum, and of this hardened [substance]," explained Villasante, "I did not make any other test or experience."[23]

Villasante tested the efficacy of the Santo Domingo balsam in Santo Domingo, in Seville, and at the court. According to his "experiences," balsam could heal wounds in a short time. It was useful for healing all types of abrasions and for relieving stomach pain. Balsam was also said to be therapeutic for the liver and gallbladder, for treating gout, and (very important) for relieving tooth pain. Villasante expected that the knowledge about his balsam would improve with time and new tests:

> This [balsam], by experience, shows already that it is beneficial for the diseases that I have mentioned. With time it may be shown by experience or reports from physicians whether it might be beneficial for other things, and they could also reveal the method for the perfection of this liquor and balsam.[24]

Villasante assumed that knowledge about balsam was cumulative and based on the experience of other physicians. This was a common tendency in the production of knowledge related to the New World. Only through new empirical information could physicians, cosmographers, and natural historians complete their study and understanding of the New World.

Villasante's report was addressed to a community of royal officials, merchants, and physicians, all of whom were learning to rely on personal experience and testing, rather than on classical traditions alone, for the production of knowledge. The crown's support for such empirical practices fostered new social relations among merchants, artisans, cosmographers, physicians, and royal officials.

Villasante, accordingly, did not place balsam within the Galenic framework of humors (blood, choler, melancholy, and phlegm) and qualities (hot, cold, dry, and moist), for instance, as the physician Garciperez Morales would do a few years later:

> Of this precious liquor, commonly called balsam, which is brought from Santo Domingo of the Indies: its first virtue is hot in the second grade, or a little less; dry in the first metha [*sic*] of the third [grade], or a little more.[25]

Garciperez Morales would write his treatise at the request of the crown in 1530. His audience was composed of both royal and regular physicians in Spain, including his student Nicolás Monardes, mentioned above (who would become well known for his research on American plants).[26] Morales framed his treatise in classical and traditional terms familiar to his audience. Villasante, by contrast, framed his account in empirical and commercial terms (as Monardes would do in his own account of American medicines and plants) and addressed his report to commercial partners and to government bureaucrats back in Spain. The difference in audiences, with their diverse interests and backgrounds, explains the difference between Villasante's and Morales's approaches to reporting on the Santo Domingo balsam. In addition, Villasante had firsthand experience of the New World and knowledge provided by indigenous people, while Morales had neither that type of experience nor that type of knowledge. The two different views show the novelty of Villasante's approach and the difficulties in integrating empirical practices in traditional approaches to knowledge. For instance, the Aristotelian tradition did not consider that personal experience constituted knowledge per se. Personal experience provided information, but only the establishment of causes constituted knowledge.

Villasante's empirical approach to nature was not new to the Spaniards in the New World. Since the mid-fifteenth century, humanists had emphasized the collection of empirical evidence to solve internal problems in their textual sources. What was new to Villasante and the Spaniards in the Indies was the intense use of empirical evidence to describe products and the institutional role played by the Spanish monarchy in this not-yet-formalized research project.

The interest of the crown in the commercialization of balsam shaped decisions not only about its production but also about the validation of empirical knowledge about it. The crown granted Villasante, his heirs, and anyone else that he deemed appropriate a complete monopoly on selling the Santo Domingo balsam as well as the other drugs he would find in the New World. Villasante also obtained in perpetuity for himself and his heirs the governorship of the fortress of Santo Domingo, Indian labor, tax exemptions, and other prerogatives.[27] All was in place for the exploitation of balsam. Other experts in the field, however, soon challenged Villasante's report.

In 1529 the Hispaniola physician Barreda first challenged Villasante. Barreda, who had been the Inquisition's physician, left Spain for Hispaniola with Pedrarias Dávila's expedition to Panama (1513–1514).

In December of 1513 the crown had approved a payment of 12,000 maravedís to Barreda for his travel expenses.[28] He held the title of royal physician in Hispaniola until 1519, when the crown suspended his title.[29] In 1526 Barreda was appointed official physician of Santo Domingo, and he probably met Villasante there.[30] But in spite of the fact that Barreda had been in the New World for almost fifteen years, it was the entrepreneur and not the physician who brought the new medicine to the attention of the crown.

Barreda argued that the crown had been deceived by the physicians in Spain who "approved as balsam the liquor that the aforementioned Villasante" took with him to Spain. Barreda claimed that the royal support for this drug would harm the person and property of the crown's subjects. He chastised Spanish physicians for failing to discuss the matter with their colleagues in Santo Domingo:

> [Spain's physicians] know or should know that they [Santo Domingo's physicians] do not lack letters, nor extensive experience, nor knowledge of the [so-called balsam] tree, its fruit and leaves, and the methods to apply the aforementioned liquor that comes from this tree.[31]

Knowledge pertaining to the New World, contended Barreda, had to be articulated by those with direct experience of the New World. Physicians in the Old World, despite their "letters," did not have this experience. For this reason, they needed to consult with their counterparts in the New World. In the interest of the well-being of the Spanish subjects, personal experience was a better source of knowledge than "letters" alone. Dr. Barreda was integrating direct personal experience into his classical training and arguing that he therefore had more authority than those without formal training or direct experience. Experience, as Villasante argued, provided the element of authority missing in those whose training only included "letters"—a significant departure from the humanist and scholastic culture that valued "letters" above all things.

In his report, Barreda noted the differences between classical balsam and the "liquor that Villasante" took to Spain.

> The main virtue of this liquor is to restrain the blood in fresh wounds by pressing it over them, and [to restrain] the flow of blood from below [the rectum], this virtue, either called opilativa . . . or constritiva . . . , in what books does it appear that balsam has this virtue?[32]

Certainly classical texts on medicine, such as the *Materia Medica* of Dioscorides, did not list this virtue among those attributed to balsam.[33] Barreda also compared the trees and the different methods used to obtain liquor from each type of tree before concluding that Villasante's liquor was not the authentic balsam. By the mid-sixteenth century scholars like Gonzalo Fernández de Oviedo, Andrés Laguna, Nicolás Monardes, Pedro Arias de Benavides, and Conrad Gesner would agree with him.[34] Nevertheless, Barreda found that this "liquor has other virtues experimented by me [*por mi spimentadas*]."[35] He noted that the Santo Domingo balsam was efficient for healing rheum as well as kidney and stomach "passion."[36]

In spite of his complaints about Spanish physicians, Barreda was not in fact alone in his criticisms. In the early 1530s, the crown complained, there were "some physicians, surgeons, and other people who, without complete information on the balsam recently discovered in our Hispaniola and without yet having made any experience with it, have published and continue to publish some publications" against it.[37] Moreover, people had decided not to buy the new balsam because of these publications, which "harms the health of the sick and wounded, and our royal treasury."[38] Such publications indicated that the criticisms had merit; but the battle for true knowledge about the balsam had just begun, and direct experience as a source of knowledge was at the center of this battle.

The crown sought to control dissident physicians by ordering them to speak or publish only after they had tested the balsam (*hecho con ello esperiençia*). Furthermore, they had to bring their findings before local magistrates, who would send them to the crown. By asking physicians to "make experiences" with balsam and then to show their reports to royal officials, the crown established empirical procedures to validate epistemological claims. Meanwhile, local magistrates were asked to foster the sale of balsam "in the best way they see fit."[39]

This situation shows the interplay between the production of new medical knowledge regarding New World medicines and the political and economic interests of the crown in controlling this knowledge and its products. In this particular case, controlling knowledge about balsam amounted to controlling the possibility of its commercialization.

The crown, however, not only attempted to press dissident physicians into experimenting with the New World balsam; it also ordered particular physicians, surgeons, and hospitals to carry out experiments with it. The crown had listened to the dissidents and sought to produce accurate knowledge. In one case, the crown sent a sample of balsam, purportedly useful to "cure injuries and many illnesses," to the Hospital of the Cardinal in Toledo for use on patients chosen by the physicians and surgeons of the hospital. The crown requested that the hospital administrators "be attentive to inform us of the cures and experiences realized in the hospital with this balsam."[40] Hospitals in Seville, Burgos, Galicia, and Granada received similar orders.[41]

The crown approached individual physicians, too. Andrés de Jodar, for instance, a resident in Baeza, received the order to use balsam for those "cures and experiences" that he would deem appropriate.[42] Moreover, he should "put in writing" whatever he might find, by means of "art" and experience, to be "certain and true," "sign" his report, and send it to Villasante's partners in Spain. Twenty-two physicians and surgeons in different cities of Spain received similar orders.[43] Villasante's partners, Franco Leardo and Pedro Benito de Basniana, would use these reports to commercialize balsam in Spain, and they hired some physicians and surgeons to help with this task.[44]

By 1532 information was already arriving at court. A certain Juan de Vargas had been using the "balsam from the Indies" to heal the sick in the area of Cuéllar (Segovia).[45] He seemed to have been quite successful, for the crown ordered the officials there to collect information from patients who had been healed with Santo Domingo balsam. The scribe of Cuéllar, Melchor de Angulo, received the information and sent it to the crown. He received 108 reales for the eighteen days he worked on this assignment.[46] The crown also asked Juan de Vargas to come to court, which he did, in late 1532 or early 1533.[47] During his stay there, he tested the balsam and received some monetary compensation for his work.[48] Despite royal support for his position, some practitioners in the medical community still opposed the use of this balsam, maintaining that it was not authentic. In 1539 the physician and apothecary of the village of Amusco (Palencia) denounced Vargas for using the New World balsam. The authorities of Amusco arrested him and took his balsam. He was later released; the crown asked the authorities to explain the matter.[49]

In the end, the crown could not dismiss Barreda's contention that the Hispaniola balsam was not authentic. The only thing that mattered, however, was that the Santo Domingo balsam, as both Villasante and Barreda had argued, was especially good for treating wounds. The crown sought, first, to develop the right method to use it; second, to end the confusion between New World balsam and classical balsam; and, finally, to convince other physicians that it was a worthy medicine.

The balsam episode highlights the emphasis placed on empirical approaches to natural products of the New World, approaches that resulted from the commercial and imperial activities of Europeans outside Europe. Certainly, entrepreneurs sought drugs that were already known to them and that were described in classical traditions; such was the case with balsam. Yet New World drugs had to be tested first; and it was the testing, not the books, that provided final knowledge about them. European notions about nature were thus adapted to incorporate discrete local settings, soon to become gardens of knowledge, into an emerging global framework of communication and trade.[50] Physicians such as Barreda and entrepreneurs such as Villasante—as well as cosmographers and pilots, natural historians, and explorers—helped to displace the supremacy of classical texts to accommodate the increasing flow of new knowledge between Spain and America.

Simultaneously, local natural settings were adapted to fit European objectives and strategies with regard to trade and the exploitation of natural resources. Contact with the New World accelerated this process of transforming and exploiting nature. Nature became a contingent reality, adaptable to human plans and needs, and a collection of commodities and curiosities ready for exploitation and collection.

GARDEN OF KNOWLEDGE

The empirical approach to nature demonstrated in the balsam case extended similarly to areas other than drugs and spices. Mercantilist policies encouraged the introduction of Old World products to the New World and thus encouraged the development of a material and empirical approach to nature. One of the most striking consequences of these policies was the transformation of the natural world of the Indies in the first fifty years of the Spanish presence.

Adapting European agricultural products to the New World was difficult because of differences in climate and geography. Ships transported seeds and cattle to the new lands, so people would grow "the same crops as . . . [in Spain], that is, wheat, barley, and grapes";[51] but they had to account for different climatic conditions.[52] The spread of animals was astonishing. In 1518 a royal official in Santo Domingo, the jurist Alonso de Zauzo (1466–1527), wrote a letter to Charles V telling him about the "marvelous multiplication of livestock." "Cows," Zauzo told the king, "give birth generally to two calves, but many times to three."[53] He mentioned that in Hispaniola there were already hogs, sheep, and mares. Zauzo suggested sending merino sheep and proposed a mechanism for doing so: "those coming first should be accustomed to eat grains and things that can be brought in a ship so they would not die."[54] Zauzo's primary responsibility in Santo Domingo was the organization of the judicial system there, but he also found the time to comment on the natural marvels of the New World. In time, Hispaniola became the source of livestock for the rest of the Indies. In 1519, for instance, the crown ordered Hispaniola officials to send a donkey to the Franciscans on the Cumana coast (Venezuela).[55]

As late as the 1650s, according to Father Bernabé Cobo (1572–1659), people were still surprised by how quickly European animals and plants flourished in and adapted to the New World. Cobo tried to explain this phenomenon. Animals and plants, he maintained, had proliferated in the New World because Spaniards took them everywhere; and, even when the Spaniards died, animals and plants survived. Cobo also observed that indigenous people liked the new plants and animals and took good care of them. Finally, he mentioned the abundance of food and water and the different natural environments that favored the increase of animals and plants.[56] Cobo was right.[57]

Adapting domestic plants and trees to the New World required more experimentation; animals multiply by themselves, but domestic plants need cultivation. More testing was necessary to adapt plants successfully. From 1494 to 1517 the crown and royal officials in the Caribbean tried many different methods for cultivating grains and trees. In 1514 a royal provision to San Juan Island ordered that its residents plant apples, pears, pomegranates, quince, peaches, walnuts, chestnuts, "and whatever grows well."[58] The crown granted tax exemptions and financial incentives to laborers to stay and work in the new lands and experiment in the planting of familiar crops. Several varieties of cereals, for instance, were tried in different parts of the Caribbean throughout the century.[59] The crown, royal officials, and ordinary immigrants shared a similar desire to transform the new lands to make them economically and ecologically habitable for Europeans.

Although the Spaniards did not know the details of this process, they knew that settlement and exploitation of the New World would have been impractical without it.[60] This transformation of nature implied experimentation and the transfer of technology to the New World as well as knowledge about the natural products specific to the Indies—the kind of knowledge that Villasante had obtained. It was a project inspired by a Christian ideology that assumed the unity of the natural world and the transformation of nature for human purposes. What was particular to and significant in the Spanish American empire was the royal support for the experts engaged in the transformation of the New World. The activities of peasants, merchants, and artisans—by themselves practical and empirical—gained in status by the support of the crown. These empirical practices were placed in larger political and economical contexts to serve the interests of the crown.

Furthermore, the transformation of nature was not limited to adapting European plants and animals. In 1534 the king ordered the governor of Tierra Firme (the area covering the coasts of present-day Venezuela, Colombia, and Panama) to take "expert people" to the Chagre River (in present Panama)—navigable from the Atlantic Ocean—and determine "what form and order could be given to open the land [between the Chagre River and the Pacific Ocean], so that, opening that land, the said river be connected to the South sea, thus resulting in navigation" from the "North sea" to the "South sea." This oceanic channel would increase commerce and thus provide a "great service" to the crown, Peru, and Tierra Firme.[61] In October the king obtained an answer from Pascual de Andagoya, governor of Tierra Firme: "I do not believe that there is a prince in the world that, with all his power, would be able to do it."[62] The practical approach to nature implicit in these activities resulted from imperial and commercial needs rather than religious attitudes. Although the idea of a canal through Tierra Firme could not be implemented at the time, Spaniards studied nature in order to reshape the natural world to fit commercial and political needs.

The Spanish crown and entrepreneurs were also interested in the introduction of commodities to the New World. From 1518 onward the crown had promoted the cultivation of mulberry trees (for the cultivation of silk), cloves, ginger, and pepper as well as dyes such as pastel and madder. In 1535 the crown, after previously unsuccessful attempts, granted a license to cultivate pastel and saffron in New Spain to two Germans, "Micer Enrique" (or "Enrique Ynger") and "Alberto Cuon." Little is known about Alberto Cuon. But Micer Enrique was Heinrich Ehinger, brother of Ambrose and Ulrich and partner of the German bankers the Welsers.[63] In the contract with the crown, the Germans obliged themselves to take to New Spain at their own expense "masters, laborers, seeds, and tools." In return, the crown granted them the monopoly in the cultivation and trading of pastel and saffron, "all the lands and people necessary" for the cultivation and manufacture of these products, and economic and financial privileges such as a tax exemption and a license to take 200 slaves to the New World.[64]

The saffron project failed because *tuzas* or *totzans* (rodents) ate the roots of the saffron.[65] The pastel project fared better, at least in the beginning. By 1537 the Germans had brought five "masters of pastel-making" from Toulouse, France, to Seville. Royal officials at the Casa de la Contratación refused the masters a license to pass, following secret instructions from Charles V to obstruct foreigners traveling to the New World.[66] When the bankers protested, the crown had to order the officials to let them pass to the New World "despite being Frenchmen."[67] The German's partner in New Spain was Alonso de Herrera, who later would organize the manufacture of beer in New Spain.[68] Production began the following year. In 1538 Alberto Cuon brought a sample of pastel from New Spain to Segovia, a famous textile center, "to test it" there, as Villasante had brought samples of balsam to test them in Spain. The queen ordered the magistrate of Segovia:

> to find out the most skillful and honest people [among dyers and other officials] of the city to test [the pastel], under oath, before your presence, and to reveal their opinions and findings, and to provide this declaration to Cuon.[69]

In New Spain, some pastel was sold for testing in 1539.

The introduction of pastel into the New World required the transference of expertise from France to New Spain and the testing of the product in different areas (Segovia and Mexico).[70] Similarly, the production of balsam had required the transfer of knowledge from Catalina de Ayahibex and her family to the Spaniards and the testing of the product in different areas (Santo Domingo and Spain). In the exploitation of natural species in the New World, the crown supported private initiatives and articulated state policies based on those initiatives (i.e., the search for new commodities). In the alliance of private and state commercial interests, natural entities became the object of empirical observations and testing that took place outside the traditional venues of guilds, universities, and humanist circles.

These empirical and testing practices became part of the American enterprise. The pastel enterprise soon failed, because it could not compete with the quality of the pastel from Toulouse. As a result, the crown became interested in the search for native dyes in the New World. In the late 1550s Marcos de Ayala, a resident of the town of Valladolid, Yucatán, presented a report on Campeche wood, a dark blue dye from Yucatán, to viceroy Don Luis de Velasco. Ayala tested this dye before the viceroy.

The viceroyal court became the place for performing "tests" in the New World. In the case of technological innovations, it was the place for legitimizing those experiences.[71] Don Luis de Velasco granted Ayala a license to exploit Campeche wood in Yucatán. In 1563 the king renewed the original license for ten years, despite Ayala's failure to exploit the product in the previous years.[72] At that time the crown was also pursuing another blue dye: indigo. Simultaneously with Ayala's report, the entrepreneur Pedro de Ledesma had sent a report on indigo to the king. Ledesma explained that the method used by the indigenous people to process indigo was costly. He requested royal support to find a cheaper way to exploit New Spain's indigo.[73]

By 1564 Ayala had begun to sell his dye.[74] In 1565 the crown dispatched a royal decree to the governor of Yucatán requesting samples of Campeche wood, indigo, and cochineal (a red dye made from the bodies of insects):

> because we want to know the qualities of this wood and seeds and understand their effects and uses in these kingdoms . . . and testing [*el ensaye*] of them as it is appropriate. We order you to send a reasonable amount of campeche wood, indigo, and cochineal to our officials of the Casa de la Contratación in the first ships leaving those provinces.[75]

By 1576 the cultivation and manufacture of Campeche wood was already an important activity in Yucatán and Campeche. Meanwhile, Pedro de Ledesma had established a partnership with the Marqués del Valle to exploit his indigo. The partnership was dissolved before 1572, however, and indigo could be grown by anyone from that time on. By 1577 there were already forty-eight indigo mills in Yucatán.[76]

The crown saw in the production of these dyes a mechanism to stabilize the prices of the textile industry in Spain and to diminish Spain's dependence on other countries. Thus, the crown requested viceroy Don Martín Enríquez to send a report about the habitat of Campeche wood and indigo, the methods of using them, plans for maintaining stable production, and the cost of production. In this way the study of nature and the welfare of the kingdoms became the double face of economic enterprises in this period.[77]

The development of commercial and imperial states during the sixteenth and seventeenth centuries fostered empirical practices because merchants and royal officials used empirical information to understand the world outside Europe, especially the New World. Merchants and royal officials gathered information about medicines, foods, climatic conditions, geography, dyes, and animals and passed that information in reports to their partners in the Old World. In some cases (for example, avocados), this information lacked references in classical traditions; in other cases (for example, balsam), the references to classical traditions did not provide actual knowledge about the new medicine. Information about rivers and the possibility of building channels had to be based on empirical information. In all of these instances, empirical information was necessary for the articulation of political and commercial decisions. The long-distance control of lands created a context for the use of empirical information as a source of knowledge, as shown in this chapter.

The long-distance control of land also created the context for institutionalization of navigational practices. At the Casa de la Contratación, the crown established several offices with the purpose of bringing the experience of pilots under the control of royal officials so that the experience would be incorporated in new charts and navigational instruments and in the training of new pilots. A center of information thus emerged within the Casa for gathering information and teaching purposes. The next chapter explores how the crown institutionalized the training of pilots and the making of instruments at the Casa's center of information.

TWO

A Chamber of Knowledge

THE CASA DE LA CONTRATACIÓN AND ITS EMPIRICAL METHODS

In 1598 the Englishman Richard Hakluyt described the navigational activities of the Casa de la Contratación in Seville for his English readers:

> [The] late Emperour Charles the fift, considering the rawnesse of his Seamen, and the manifolde shipwracks which they susteyned in passing and repassing betweene Spaine and the West Indies, with an high reach and great foresight, established no onely a Pilote Major, for the examination of such as sought to take charge of ships in that voyage, but also founded a notable Lecture of the Art of Navigation, which is read to this day in the Contractation house at Sivil. The readers of which Lecture have not only carefully taught and instructed the Spanish Mariners by word of mouth, but also have published sundry exact and worthy treatises concerning Marine causes, for the direction and incouragement of posteritie. The learned works of three of which readers, namely of Alonso de Chavez, of Hieronymo de Chavez, and of Roderigo Zamorano came long ago very happily to my hands, together with the straight and severe examining of all such Masters as desire to take charge for the West Indies.[1]

Although the office of the chief pilot was established by Ferdinand the Catholic and not by Charles V, Hakluyt's description captures the teaching and training activities of the "Contractation house at Sivil." The office of the chief pilot, however, was established not only to train pilots but also to make charts. The cosmographers of the Casa were hired to make instruments and lecture pilots. The institutionalization of these activities was the result of the Spanish crown's political interests and desire for control in the New World. In the process of gaining control of long-distance lands, the crown established mechanisms at the Casa for making navigational instruments and charts, for teaching navigational techniques to pilots, and for examining the pilots. The establishment of these mechanisms (the focus of this chapter) was the result of administrative practices copied from the Portuguese, suggestions made by pilots and cosmographers, and administrative decisions for solving disputes among pilots and cosmographers working at the Casa.

The overarching goal of the Casa was "to bring together practice and theory," which was a way of bringing together the personal experience of pilots and the formal education of cosmographers.[2] Personal experience was the key element in the collection of new knowledge but had to be organized within a theoretical framework to be useful. Thus, the cosmographers at the Casa used that knowledge to create charts and to train pilots in the use of instruments (so they could use those charts). During the entire sixteenth century, the tensions between these two groups were significant and constant, and they constitute a central theme in this chapter. The crown would intervene often to mediate between the two groups, acknowledging the need for bringing together practice and theory and the difficulties in accommodating those approaches in a coherent practice.

In the early sixteenth century European kingdoms sought to establish areas of influence over routes to Africa, Asia, and the New World. Portugal and Spain claimed rights over different areas of the Atlantic Ocean, yet they first had to determine the size of the ocean, the exact locations of the recently encountered lands, navigational routes, and geographical aspects of the new lands. A set of offices within the Casa emerged to face the challenge of collecting and organizing knowledge of the new lands into maps, charts, instruments, treatises, and navigational practices.

A veritable chamber of knowledge, a center of information, evolved as the offices and practices of the Casa were developed and institutionalized to collect and disseminate information about the New World, to train artisans (pilots) in the new navigational techniques, and to hire professionals (cosmographers and pilots) for research and teaching activities. The historian Clarence Haring calls it a "Hydrographic Bureau and School of Navigation, the earliest and most important in the history of modern Europe."[3] The Casa's activities predated Francis Bacon's (1561–1626) program of research and the information-gathering activities of scientific academies such as the

Royal Society of London for Improving Natural Knowledge (1660). Not surprisingly, Bacon's program and the activities of the Royal Society emerged at the time when England began serious expansion into the New World and some seventy years after the publication of English translations of the most important Spanish books on natural history, navigation, and medicine related to the New World and the Casa.[4] Long-distance empires and empirical institutionalized practices emerged during the sixteenth and seventeenth centuries as the result of the European expansion into the Atlantic world.

This center of information emerged slowly from the interaction between royal officials and pilots, explorers, and artisans engaged in enterprise in America. In this world of exploration, knowledge became a central element in imperial politics and commercial activities. Its gradual creation mirrored the establishment of the Spanish kingdoms in America. By 1550 the Spanish had established towns and cities from Mexico to Chile and had already transplanted Old World products to the New World. While the conquest of the New World relied on force and violence as well as luck and determination, it also relied on knowledge, as the search for commodities and the writing of reports testified. Consequently, by the 1550s the Casa had already institutionalized practices for gathering, disseminating, and classifying knowledge and for training, certifying, and hiring lay practitioners.

The 1550s marked a shift in Spanish activities in the New World but also in Europe. Philip II married Mary Tudor (1554), adding European Atlantic interests to Spain's existing Mediterranean ones, which continued even after the death of Mary in 1558, with the war in the Netherlands (1560s) and the annexation of Portugal (1580). In 1559 Spain and France signed the treaty of Cateau-Cambrésis, ending more than forty years of war and beginning a new alliance to combat the Protestant advances in both countries and outside. The treaty also finally gave Spain control over Italy and the Mediterranean routes. The ideological tensions between Protestants and Catholics in Europe were beginning to shape into a "hot war," with significant consequences for international as well as domestic policies.[5]

In the New World, the 1550s marked a shift from conquest to consolidation. This is the period when las Casas took his case for the defense of the Native Americans to Spain. His efforts did not stop the process of conquest and colonization, but his arguments in favor of the Native Americans shaped later Spanish policies that sought to regulate the process of colonization more effectively.[6] In the 1550s the amalgamation process, a more efficient method to extract silver, was developed in New

Spain and introduced later in Potosí. This method helped to increase the amount of bullion exported to Spain, and the rest of Europe, from the 1550s onward. In 1552 the crown established the fleet system to protect commerce between Spain and the New World. In 1554 the archbishop Alonso de Montúfar arrived in Mexico with inquisitorial powers and began the persecution of Protestants in Mexico, Guatemala, Nicaragua, and Yucatán.[7] These events in the Old and New Worlds were structurally correlated, not causally related. The legal, economic, and religious power of the Spanish crown in the New World as well as in the Old World increased after the 1550s. This increased power was reflected as well in the number of navigational offices established in the Casa by the 1550s.

The development of the Casa during the sixteenth century was not only the result of commercial and political expansion and consolidation but also the result of particular Iberian and Mediterranean traditions, especially in the realm of navigational and cosmographical practices and techniques. Methods of navigation in the Mediterranean had been developed by traveling close to coasts.[8] Pilots calculated their position by means of an imaginary line running from east to west, using their knowledge of the winds and the type of ship as well as an hourglass. A pilot's assessment of these factors was used to calculate the position of the ship—always adding more leagues (nautical miles) to be on the safe side. As the ship approached land, a watchman would prevent it from crashing.[9] This situation changed when Iberian pilots ventured into the Atlantic, which lacked the navigational markers of the Mediterranean Sea; so pilots learned to rely instead on astronomical markers.

PORTUGUESE PRECEDENTS

Iberian navigational and cosmographical traditions emerged from the transformation of Arabic theoretical astronomy into practical knowledge and the interplay between this practical knowledge and economic and religious interests. Portugal sent the first explorers to the Atlantic as an attempt to break into the trade routes of the Muslims by way of West Africa. Spain, once it had won the wars of the Reconquista, would attempt to do the same by going west.

The Portuguese developed the most important technological knowledge for safe navigation in open seas during the fifteenth century. This knowledge supported imperial and commercial interests in opening safe routes to the East and resulted in the establishment of Portuguese routes around Africa and Spanish routes westward to the East.

At the Spanish court of Alfonso X el Sabio (Alphonse X the Learned, r. 1252–1284), scholars endeavored to preserve classical and Arab cosmographical traditions.[10] There Gerard of Cremona translated Ptolemy's *Almagest* from Arabic (the language in which most classical texts entered Europe during the Middle Ages) into Latin in 1175, together with the critical work of Jabir ibn Aflah on Ptolemy; John of Seville translated the work of al-Farghani (Alfraganus) on Ptolemy around 1134. These works survived in the so-called *Libros de astronomía* of Alfonso X; they contained a great deal of the cosmographical knowledge that the European explorers would later use to navigate the Atlantic.

The *Libros de astronomía* discussed astronomical matters such as the determination of latitude based on the sun's distance from the horizon and the construction and use of flat and spherical astrolabes, quadrants, clocks, and astronomical tables.[11] The Portuguese transformed these data into practical knowledge, thereby linking developments in navigational astronomy with the growth of overseas expansion. The Portuguese experience in the Atlantic would in turn provide the basis for Spanish navigational institutions and practices in the early sixteenth century.

A number of factors prompted the Portuguese to transform classical knowledge into practical knowledge. They had economic and religious interests in exploring the African coast; they built ships equipped to navigate the Atlantic; and, finally, they developed an atlas of constellations to help Portuguese sailors find their positions while far out at sea. Northern and Mediterranean traditions of shipbuilding and cosmographical research by Iberian Jews influenced shipbuilding and navigation techniques in Portugal. Although the *Libros de astronomía* provided instructions for determining the latitude of a given place, they did not explain how to apply latitude to navigation.

The work of the exiled Spanish Jew Abraham Zacuto (ca. 1452–1515/1522) and the support he received from the Portuguese crown proved crucial to the development of navigational techniques later on. Zacuto was born in Salamanca in 1452. He studied under the direction of Rabbi Ishaq Campanton and Rabbi Ishaq Aboab and may have studied at the University of Salamanca. In Salamanca, under the protection of Bishop Gonzalo de Vivero, he taught mathematics. Later he moved into the service of Juan de Zúñiga, official of the Order of Alcántara, in Extremadura. Zacuto worked in Spain until the expulsion of the Jews in 1492. He then moved to Lisbon,

Portugal, under the service first of King John II (1481–1495) and then, after John II's death, of Manuel I (1495–1521). In the opinion of scholar David Romano, Zacuto "was the last important practitioner of Spanish Jewish science on the Iberian Peninsula."[12]

Zacuto's work introduced the novel method of using solar observations to determine latitude. This innovation emerged from the practices developed in Portugal during the reign of John II. Solar observation methods used the highest point of the sun above the horizon to determine the altitude. The sun's altitude varies throughout the year; so pilots needed tables of solar declination, charting the position of the sun month to month. Zacuto's tables constituted the basis of navigational regiments (tables of solar declination) that he created along with his disciple José Vizinho. These regiments became the first practical manuals used by the Portuguese sailors in their navigation around Africa.[13] Of course this system worked only during the day; at night pilots needed to determine latitude using the stars.[14] For this, they used the atlas of constellations composed by Abraham Cresques, a Jew from Majorca.

With these tools, Portuguese cosmographers developed charts based on the altitude of the sun (during the day) or a pole star (at night) above the horizon. If they wanted to find their latitude in any given place on the Atlantic Ocean, along the African coast, they would find the altitude of the sun or polar star at that place. By subtracting that height from the altitude of the sun or polar star in Lisbon, they would calculate their distance south from Lisbon. This new technique of navigation basically compared the altitude of a given celestial body (a star or the sun) at the point of departure (e.g., Lisbon) with the altitude of the same celestial body at different northerly or southerly points along the voyage.[15]

This new method prompted the simplification of cosmographical instruments (such as astrolabes and quadrants) for Atlantic navigation. Instruments designed to take measurements on solid earth were difficult to use on a mobile vessel. The Portuguese simplified astronomical instruments as well as the tables of solar declination. The mariner's astrolabe, for instance, was used to plot the sun near the meridian. It developed from the astronomer's planispheric astrolabe but is much heavier (for steadiness) and has openings to reduce wind resistance.[16] They were also pioneers in the introduction of astronomical classes for pilots and sailors.[17] Sailors from Spain and later Europe learned from the Portuguese to navigate with the stars. Christopher Columbus, for one, knew about these navigational techniques; he even annotated his Spanish translation of Zacuto's *Almanach* before he set sail on his famous failed voyage to the East.[18] Spain, and especially the Casa, became the repository of this technique of navigation—with the type of knowledge, instruments, and professionals necessary for its realization.

Finally, a new type of vessel, the caravel, crystallized the commercial and technical interests of the Portuguese crown and made possible the actual transportation of human and material cargo as well as ideas and practices.[19] In the Mediterranean, winds came from behind the ships; in the Atlantic, ships faced contrary winds or no winds at all. The caravel was designed to sail in shallow waters or in high seas and employed lateen (triangular sails) together with square sails. Lateen sails allowed ships to sail with contrary winds in a zigzag pattern called tacking. The Portuguese learned to calculate distances when tacking. This calculation involved both geometry and plane trigonometry, and sailors used special tables called *toletas.*[20] The Portuguese thus developed the most important technological knowledge and instruments to navigate safely in open sea. These Arab, Mediterranean, and Portuguese traditions of navigation took root in Spain at the Casa de la Contratación as Spain established its empire in the New World.

A CHAMBER OF KNOWLEDGE

On February 14, 1503, Isabel and Ferdinand ordered crown officials to establish in the dockyards of Seville a "house for trading and negotiation with the Indies, Canary Islands, our other discovered islands, and would-be discovered islands."[21] The idea may have come from a 1503 report, attributed to Francisco Pinelo, proposing the establishment of a warehouse for the centralization of trading with the Indies—and surely the Portuguese Casa da India (before 1503 called Casa de Mina e da India) was the model.[22] The Portuguese Casa "included an organization equivalent to a modern hydrographic office, at whose head was a cosmographer-in-chief. He was assisted by cosmographers whose business it was to draw and to correct charts and to compile books of sailing directions" and perhaps "to assist in the instruction of pilots," as did Spain's Casa later.[23] But the Portuguese chief cosmographer's office was at the court, not at the Casa, and he had access to information from mariners and pilots who came to the Casa da India. The teaching, in the Portuguese case, also seems to have taken place at the court.[24]

The royal decree of 1503 called for the appointment of a factor (a revenue collector for the crown), a treasurer, and a clerk to deal with the administration of commerce and fleets between Spain and the Indies.[25] The queen ordered the building of an actual "house"; but a few months later she changed her mind and ordered officials to set up the "house" in the Old Alcazar (castle) of Seville. The name stuck, though, and the rooms occupied by the officials became known as the Casa de la Contratación. The economic situation of Seville helps to explain this change in plans: from 1503 until 1508 famine and plague hit Seville hard.[26] That the house went ahead at all in such difficult times is a sign of the project's importance.

The structure of the Casa emerged slowly over the years, office by office. New officials joined the three original ones: a chief pilot (1508); a ship inspector (1518); a cosmographer (1523—to make instruments for navigation); a representative of the Casa in Cádiz (1535); a fiscal lawyer (1546—to protect the interest of the royal treasury); another cosmographer (1552—to teach cosmography); a legal advisor (1553); a president (1579); and two official lawyers (1583; a third lawyer was appointed in 1593). The lawyers formed the newly established chamber of justice (1583); the other officials remained as members of the chamber of government. In 1588 the crown appointed a purveyor, in 1612 a treasurer; and in 1625 Philip IV appointed Conde Duque de Olivares life magistrate to the Casa.[27]

By the late sixteenth century, the structure of the Casa consisted of the chambers of government and justice and their respective officials. The first three officials of the Casa initially lived in the same building; in 1518, however, Charles V ordered them to move out and use the Casa only for official meetings and business.[28] He later ordered them to live at the Casa again.[29] The Casa regulated and supervised commerce and business with the New World, payment of taxes and tariffs, and civil and criminal matters resulting from these activities.[30] The navigational and cosmographical offices belonged to the chamber of government.

These offices emerged, as did the Casa itself, from the responses of the crown to the challenges of the New World. On March 22, 1508, Ferdinand appointed the first chief pilot—initially, for organizing nautical information but soon for examining pilots in the use of instruments as well.[31] Both activities were indispensable for mastering the westward route to the East and ensuring safer trips and more secure cargoes.

By the 1550s the Casa already provided salaried offices for research (focusing on the creation of royal sea-charts and the making of instruments), for sharing and disseminating that new information and instrumentation, for licensing pilots, and for monitoring the availability of jobs for those certified.[32] Thus, the development of knowledge-producing structures, practices, and professionals was bound up in the building of empire. How this process of institutionalization occurred is quite revealing with regard to the means by which scientific knowledge became part of the expansion and solidification of empire.

In the next sections, I discuss the process of establishing offices at the Casa for the dissemination of Atlantic navigational methods. These offices offered the opportunity to hire and train pilots in these methods, which combined experience in navigation and formal training—the theory and practice of navigation. Attempting to reach a balance between these areas, the crown and its officials would usually hire different people with experience in navigation (the pilots) and formal knowledge in cosmographical matters (the cosmographers).

The empirical aspects would be particularly important in the making of charts and examination of pilots, and the cosmographical aspects would be particularly important in the training of pilots and the making of instruments. The Casa, then, hired experts in navigational matters and thus created a space for collecting and disseminating (through charts and training of pilots) knowledge about the New World. In other words, the Casa provided a space for, and thus validated, personal knowledge as a source of knowledge.

In the final section of this chapter, I discuss a report about the Casa from the 1570s. This report shows that many pilots and several Casa officials emphasized personal experience as the key element in the training of pilots and that the relationship between experience and formal training (or formal education) was constantly redefined by different groups according to their own agendas. This case shows how pilots and experts were negotiating the uses of empirical information in the context of the empire's interests. The navigational center at the Casa became the place to normalize and institutionalize knowledge and practices by creating offices and by allowing for the negotiation of empirical and theoretical practices in the production of knowledge.

EXAMINING PILOTS

In 1506 Ferdinand returned to Spain as king, after Philip I's short reign (1506). Juana, Ferdinand and Isabel's daughter, had inherited the crown of Castile from Isabel; but after her husband Philip I's death, she had become mad and incompetent to govern. Soon after arriving in Spain, Ferdinand reorganized the administration of Castile's holdings in the Indies. In addition to appointing the bishop Juan Rodríguez de Fonseca (1451–1524) and the royal secretary Lope Conchillos to oversee the general administration of the Indies, Ferdinand appointed Amerigo Vespucci (1454–1512) as the first chief pilot of the Casa on March 22, 1508.[33]

The office of the chief pilot was established after King Ferdinand called to court the pilots Juan Díaz de Solís (ca. 1470–1516), Vicente Yáñez Pinzón, Juan de la Cosa (d. 1510), and Amerigo Vespucci to organize new discoveries in the Indies, for these "had been neglected during his absence."[34] In this meeting Amerigo Vespucci "was chosen to remain at Seville to draw sea-charts, with the title of *piloto mayor* [chief pilot]."[35] America is thus named after the first pilot who was officially engaged in charting routes of access to the New World.[36] (See Appendix 1, Table 1, for a list of pilots.) The king stated the situation in these terms:

> It has come to our knowledge and it is known from experience that pilots are neither as expert nor as trained as they should be to steer and govern the ships under their command in their voyages to the Indies, islands, and Tierra Firme of the Ocean Sea. The pilots' lack of knowledge to steer and govern [their ships], to use the quadrant and astrolabe, and to take the altitude has caused and still causes many mistakes and faults in their navigation. This causes a great disservice to us and great harm to the merchants of the Indies.[37]

The duties of chief pilot included the examination and approval of pilots as well as the elaboration of the master sea-chart, called a royal portolan, from which particular sea-charts had to be drawn.[38] These functions defined the kind of knowledge the Casa would develop and emphasized the transformation of pilots into efficient imperial agents through their training and examination. Over time these functions would be ascribed to the Casa directly, as an institution, rather than to the chief pilot as a person.

Pilots' lack of knowledge in the use of instruments caused "many ill services to us [Ferdinand] and great harm to the merchants of the Indies."[39] Vespucci himself experienced the dangers of pilots' ignorance in cosmography. During one of his voyages, in 1501, his ship was lost after a storm, and none of the pilots or mariners knew where they were for some fifty leagues. Vespucci took out his quadrant and astrolabe and found their position.[40] In 1512 the crown ordered Vespucci to examine pilots in the use of the quadrant and astrolabe (see Figures 1 and 2) and to instruct them so that pilots could "bring together the practice and theory [*junta la platica con la teoria*]" of navigation.[41] When Vespucci died, his successors, Juan Díaz de Solís and Juan Vespucci, continued his practices

Parte.

da y aſtrolabio: y alapunta tenga vn agujero prolõgado adõde quepa vna chaueta q̃ apriete el halhidada conel aſtrolabio, de manera q̃ el alidada pueda ãdar al rededor del aſtrolabio como pareſce en la preſẽte figura.

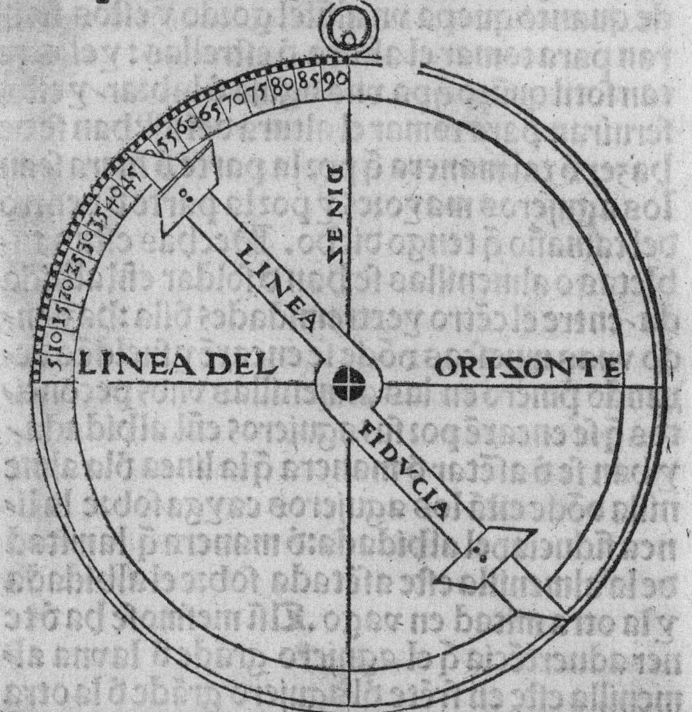

Para tomar el altura d̃l ſol, cuelga el aſtrolabio dela armilla y põ el alhidada cõtra el ſol, y alçala o baxala enel quarto graduado haſta q̃ el rayo d̃l ſol entre por el agujero pequeño d̃ la pinola, y de preciſſo eñl otro agujero peq̃ño de la otra pinola: y entonces miraras la linea fiducia y quãtos grados ſeñalare eñl q̃rto graduado començando dela linea del orizõte, tantos

1. *Astrolabe in Martín Cortés,* Breve compendio de la sphera *(Toledo, 1551). Reproduced courtesy of the John Carter Brown Library at Brown University.*

at the Casa. Pilots who were not examined and approved by the chief pilot could no longer work. Originally the chief pilot performed his duties alone at his house; but this personal system of diffusion and regulation of knowledge would change in the 1530s, during the tenure of the Italian Sebastian Cabot (ca. 1480–1557) as chief pilot (1518–1548).[42]

Sebastian Cabot came to Burgos, Spain, in 1512, in the service of Lord Robert Willoughby of Broke. Cabot met with Rodríguez de Fonseca and Lope Conchillos in Burgos to talk about "navigation to the Indies." He knew about English expeditions to North America in the 1490s and early 1500s and had experience making charts.[43] The royal counselors mentioned Cabot's knowledge to King Ferdinand, who subsequently asked Lord Willoughby to let Cabot come to his court.[44] In 1512 the king appointed Cabot captain of "the things of the sea."[45] Before moving to Seville, Cabot went back to England to close his house and collect his family. Henry VIII did not seem to be very supportive of new explorations, and Cabot would go back to England after the death of the king in 1547. By 1514 he was back in Spain. In 1518 Charles V appointed him chief pilot: he held the office for thirty years, until he "deserted the Spanish service for the pay of England."[46] But he had already done something similar in 1512—artisans were on the move, looking for the best jobs around.

In the mid-1520s Cabot was ordered to examine pilots "who would be responsible for the ships navigating our [the king's] oceans of the Indies and the Ocean Sea [the Atlantic]" in the use of instruments and routes at his house before all "the pilots present in Seville by the day and time" of the examination. The pilots had to be natives of the kingdom of Castile. Foreigners needed a royal dispensation in order to be examined or allowed to have charts or "paintings" of the Indies. The examinee had to bring proof (documents or a witness) that he had sailed for six or more years to the New World and that he had been in Hispaniola, Cuba, Tierra Firme, and New Spain. He was required to bring his own chart, astrolabe, and quadrant and demonstrate that he could use them and to answer the questions posed by the chief pilot and pilots. They had to "ask the best and more difficult questions" possible and would vote "freely without taking into consideration hatred or friendship or any other passion."[47]

A competent pilot thus was defined as one with direct experience of the New World as well as the knowledge and literacy to use and read instruments and charts—able to answer theoretical as well as practical questions about navigation and cosmography from his peers. The certification in effect institutionalized practical knowledge associated with experience, literacy, and instruments and created a group of experts who validated the

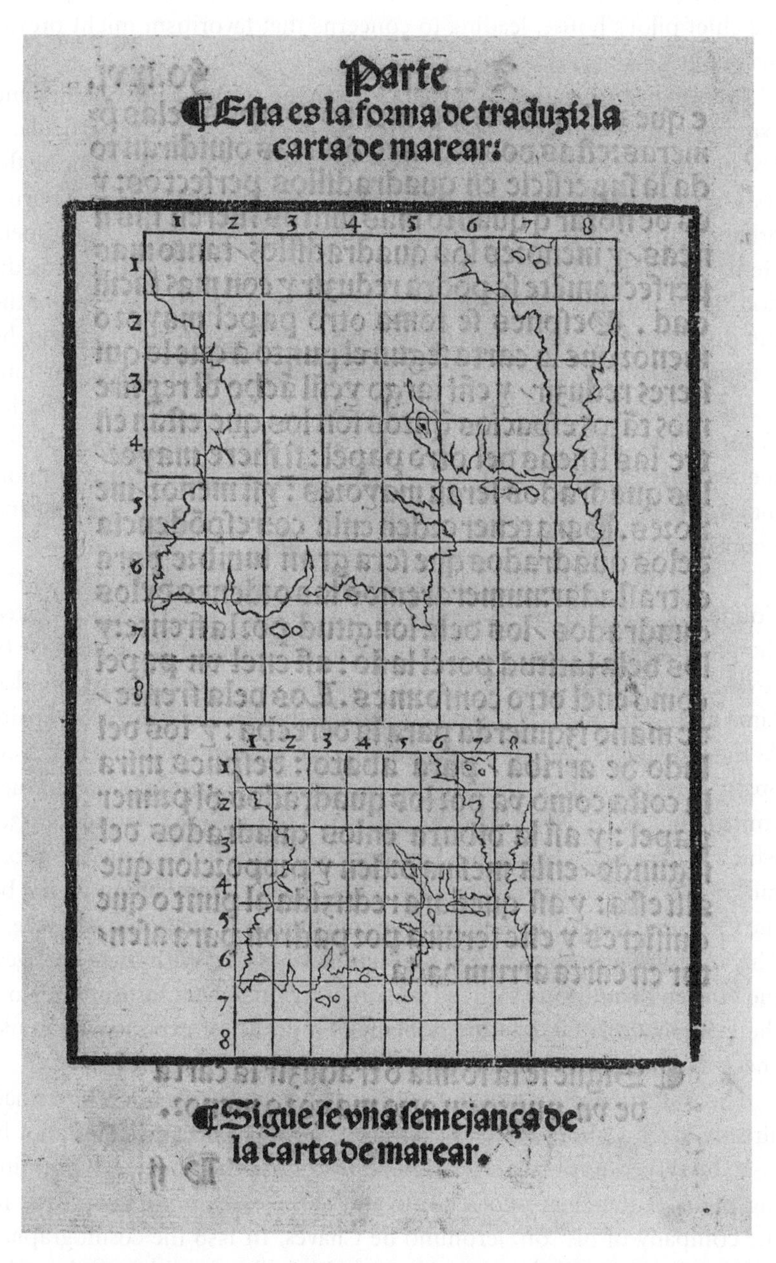

2. *Navigational chart in Martín Cortés,* Breve compendio de la sphera *(Toledo, 1551). Reproduced courtesy of the John Carter Brown Library at Brown University.*

knowledge of individual pilots. Yet the certification process took place at the chief pilot's house, leading to concerns that favoritism might prevail over ability.

Eventually, Alonso de Chaves took over the duties of instructing and examining the pilots, because Cabot left Spain in 1526, during a particularly hot summer, as chief pilot of an expedition to the Moluccas.[48] Among the sixty-seven merchants who financed his expedition was Robert Thorne, originally from Bristol and now a resident of Seville. Two Englishmen, Henry Latimer and Roger Barlow, accompanied Cabot.[49] Cabot's expedition never reached the Moluccas; instead, Cabot decided to explore the Río de la Plata in South America. When he returned to Spain in 1530, he was put in jail and later (1532) condemned to exile in Morocco for four years. "But Cabot must have an extraordinary relationship with Charles V, for in the same year the emperor recalled him from exile, canceled the sentence, forgave the fines, and reinstalled him in the office of pilot-major."[50] Chaves, however, continued with the duties of instructing and examining pilots.

In 1535 the royal official Juan Suárez de Carvajal, a member of the Council of Indies (1528–1542) and later bishop of Lugo (1539), inspected the practices of the Casa. The inspection correlated with the increase of royal control in the New World: that same year, Francisco Pizarro founded Lima and Antonio de Mendoza established the viceroyalty of New Spain. As the crown was extending its power in Peru and consolidating it in New Spain, it was also strengthening its institutions at home for accessing and controlling the New World. Suárez's visit led to the Casa's statutes of 1536, which ordered the chief pilot to examine sailors in front of the Casa cosmographers and other pilots.[51] The place for the examination would be the Casa, and the cosmographers present at the examination could vote.[52] This measure was reinstated in the statutes of 1552.[53] With these changes, the burden of authority shifted from an individual to an institution. Now the examinations became the domain of a group of experts who could check and balance the authority of the chief pilot.

The transition did not occur without tensions. Alonso de Chaves had already taken Cabot's place as examiner. When he became chief pilot in 1552 (he served until 1586), he decided to restore the practice of examining pilots at the chief pilot's house and alone—or, in his own case, in the company of his son, Jerónimo de Chaves. In 1556 the cosmographer Sancho Gutiérrez filed suit against the chief pilot for this practice. The Casa de la Contratación resolved the matter in favor of Gutiérrez and ordered Chaves to return to public examinations.[54]

In addition to changing the place and structure of the examination, the crown soon found it necessary to change its content, specifically in the use of instruments and charts. The need for more and better instruments led the crown to appoint artisans to make and sell instruments at the Casa. Various royal technicians made different instruments, which led to conflicts. The crown finally decided to intervene and established a single criterion for making instruments. This event in turn led to the creation of a lecture of cosmography, because the criteria established by the crown for making instruments assumed a level of cosmographical knowledge that the pilots lacked. The next two sections of this chapter explore these changes.

MAKING INSTRUMENTS

In 1523, a year after the return of Charles V from the Holy Roman Empire, the crown established the office of the instrument-making cosmographer and appointed to it the Portuguese Diego Ribeiro (or Ribero).[55] Ribeiro had gone to Spain in 1519 in the company of the cosmographer Ruy Faleiro to join Fernando de Magellan's expedition.[56] Ribeiro, who accompanied Vasco de Gama on his voyage to the East in 1502, worked with Ruy Faleiro making charts and instruments for Magellan's voyage. Eventually, Magellan would take with him six pairs of compasses, twenty-one wooden quadrants, six metal astrolabes and one of wood, thirty-five compasses, four big boxes for four compasses, eighteen sandglasses (hourglasses), and parchment charts made by Nuño García—one of them was a globe.[57] In 1520 Ruy Faleiro fell sick and went back to Portugal to rest. There he was arrested by order of John III for passing information to the Spaniards.[58] Ribeiro continued his work with Pedro Reinel and his son Jorge, who worked for both the Spanish and Portuguese crowns.[59] After Magellan's fleet departed, Ribeiro remained in Seville and was appointed cosmographer for making instruments and charts in 1519.

Ribeiro also participated in the Junta of Badajoz in 1524, established to discuss the matter of the demarcation line with Portugal set by the Treaty of Tordesillas in 1494 (in which Spain and Portugal agreed to divide the world between them), and in the organization of García Joffre de Loaísa's expedition to claim the Moluccas for Spain, departing from La Coruña.[60] His activities included making charts and instruments as well as designing a bilge pump. Ribeiro, in sum, was a well-trained expert in navigational matters, with practical experience and cosmographical knowledge.

Before appointing Ribeiro as cosmographer for making instruments and charts, the crown had already addressed the need for instruments by appointing Nuño García de Torreño (d. 1526) as master of instruments in 1519. He made thirty-two charts for Magellan's voyage.[61] In 1524 the crown appointed a second master of instruments, Jorge Reinel.[62] The manufacture of instruments became an important business in Seville. Pilots, according to the statutes of 1536, had to buy their own instruments of navigation before leaving Seville. A good working relationship between the chief pilot and the cosmographer of instruments meant that the Casa controlled the market of instruments in Seville, not only in terms of money but also in terms of the kinds of instruments in use. This market dimension pushed the crown to establish legislation to regulate the type of instruments and to establish a lectureship in cosmography.

Another factor in the crown's decision to regulate instruments was a dispute about chartmaking between the cosmographers Pedro de Medina and Diego Gutiérrez. Originally, Gutiérrez and Cabot dominated the market of charts and instruments in Seville; Gutiérrez made the instruments and charts, and Cabot approved them. Diego Gutiérrez (ca. 1486–1554) had been in Seville ever since the time of the Magellan expedition; it is possible that he began working with Nuño García de Torreño. Gutiérrez received a royal license to make charts and instruments in 1534. His son Sancho received a license in 1539, and it seems that two other sons, Diego and Luis, were also involved in manufacturing instruments. Over the years, the Gutiérrez family established close connections with the chief pilot, Sebastian Cabot, ensuring virtual control of the instrument-making business in Seville.[63]

When Medina arrived, he received a royal license (in 1538) and began selling his own instruments and charts. His charts were different from those made by Gutiérrez and approved by Cabot. At issue was the authenticity of their charts. Gutiérrez made his according to the traditional ways of navigation, using two systems of gradations (latitudes) in the same chart to correct for the magnetic alteration of the compass in the Atlantic. Medina instead was following the portolan, which used the same system of grades and distances for all regions.

In 1545 the Council of Indies decided in favor of Medina and asked Gutiérrez to make his charts according to the portolan. The Casa officials then wrote to the council and explained that pilots would now have to learn how to navigate with the new charts and must thus all be retrained—a huge effort. The Casa's activities from the examination of pilots to the making of instruments could only produce results if the pilots had a basic knowledge of long-distance navigation. The crown sought to transform the experience of the pilots into useful knowledge to make the voyages to the New World secure and efficient.

THE CHAIR OF COSMOGRAPHY

In response to the Casa's complaints, the crown established the chair of cosmography in 1552 and appointed Jerónimo de Chaves (1523–1574) to fill the position, which he did until his death. When the Casa was founded, the chief pilot taught the theory of navigation to any pilots who wished to learn it. But the number of pilots with theoretical knowledge and expertise in the use of instruments did not increase rapidly enough to serve the number of voyages to the New World, which generated tensions between pilots and royal officials throughout the sixteenth century. The administrative solution to these tensions would, in turn, shape the teaching and practice of navigation during this period.

Cosmographers and mathematicians soon offered their instructional services to the crown. Once again, the initiative to establish a salaried office to teach cosmography to pilots came from below. For instance, in 1525 Fray Juan Caro—a Spanish Dominican working for the Portuguese crown in Cochin, India—wrote to his relatives in Seville and to Charles V, offering his services and requesting a paid position to teach Spanish pilots the art of navigation in Seville. Caro, who had traveled to India to learn the "secrets [of navigation], so with those and his knowledge he could receive honors" in Spain, "wanted very much to live [again]" in Castile, his native country; however, he was ordered to his monastery in Portugal. There the Portuguese king ordered his arrest—his knowledge of the Moluccas made him dangerous.

Caro was exiled to Sofala, Africa, where he died. It should be noted that Caro's proposal for teaching pilots came before the crown had established a lectureship of cosmography separate from the teaching duties of the chief pilot.[64] Others closer to the crown also suggested the establishment of this office, especially during the absence of the chief pilot, Sebastian Cabot, when Alonso de Chaves taught the theory of navigation and use of the astrolabe, quadrant, and navigational charts at the house of Hernando Columbus. These lectures were the result of the initiative of Columbus and Chaves.

Columbus's participation in this initiative was important because he was well known and respected at the court. Although he was not an official member of the Casa, he had already assisted the crown in various cosmographical activities. In 1517, for example, he had begun a "cosmography and description of all Spain,"[65] by collecting, through messengers, "all the peculiarities and memorable things that there are [in Spain]."[66] In 1523, for reasons that are still unknown, Charles V eliminated the project.[67] Still, the crown soon requested Columbus's services again, asking him to join the Junta of Badajoz (1524).[68] In 1526 the crown ordered Columbus to make a master portolan, a mappa mundi, and a globe representing the new lands.[69] In 1528, when Chaves was already instructing pilots at Columbus's house, Columbus recommended that the council hire Chaves to teach pilots at the Casa. This recommendation constituted the first attempt to separate the instruction of pilots from the office of the chief pilot. This recommendation was an attempt to create new salaried positions for those trained in the new navigational techniques as teachers and thus to create institutions for those willing to learn these new navigational practices and support them.

Although the council did not follow Columbus's recommendation, it thanked Chaves and asked him to continue instructing pilots.[70] Officially, the chief pilot continued with the instruction of pilots. The creation of a separate office for teaching pilots would not occur until after Cabot ceased to be the chief pilot (in 1549) and until the crown was pushed into a decision by the dispute between Pedro de Medina and Diego de Gutiérrez. When the crown appointed Alonso de Chaves as chief pilot in 1552 (following the temporary appointment of Diego Sánchez Colchero from 1550 to 1552), it also established the lectureship of cosmography and appointed to that post Chaves's son Jerónimo.[71]

Jerónimo de Chaves was born in Seville in 1523 and attended the university there.[72] He obtained a bachelor's degree in mathematics and perhaps studied medicine. In 1545 he published a Spanish translation of Johannes de Sacrobosco's *Tractatus de Sphaera Mundi,* a popular cosmographical treatise that explains Ptolemy's spherical geometry.[73] Chaves intended to make Sacrobosco's popular cosmography available to practical astronomers ignorant of Latin. In 1548 he published a chronology and almanac.[74] This work proved very popular and went through some eighteen editions by 1588.[75] Chaves was also a collector of natural curiosities, books, maps, and instruments. His appointment acknowledged his growing status within the community of cosmographers at Seville. Pilots could no longer take their examinations "without listening for a year to that science or most of it to gain competence."[76]

Jerónimo de Chaves received instructions from the king specifying the duties of his professorship. These instructions coincided with the statutes of 1552. According to the new rules, studying cosmography and the new navigational methods was no longer voluntary for pilots. The mandatory program consisted of a year of lectures covering Sacrobosco's treatise on the sphere and the magnetic variations of the compass, using the Spanish translation of the treatise by Chaves; the use of regiments; the use of instruments for taking altitudes such as the astrolabe, quadrant (an instrument in the shape of a quarter-circle for making angular measurements), and cross-staff; the use of the compass; and the use of diurnal and nocturnal clocks. In addition, pilots had to memorize the lunar calendar to know the time of tides. Classes took place Monday through Friday from three to four o'clock in the afternoon from October to March and from five to six from April to September.[77] A year proved to be a long period for the pilots to be away from their jobs at sea, and Jerónimo de Chaves soon asked the Council of Indies to reduce the lectureship of cosmography to three months.

With the addition of the chair of cosmography and the statutes of 1552, the structure of the Casa's chamber of knowledge was complete (see Appendix 1, Tables 2 and 3). The statutes of 1552 codified the following practices: examination of pilots had to take place only at the Casa, before royal cosmographers and regular pilots, and certification was granted by a majority of votes from those present at the examination (number 128). Examinees needed to have six years of practice in navigation and knowledge of the theory of navigation (number 129). Furthermore, the chief pilot could neither instruct nor sell instruments and charts to prospective examinees, so that no conflict of interest could arise (numbers 130 and 131). Finally, the chief pilot, cosmographers, and pilots were entrusted with the crucial task of emending the royal sea-chart together (number 134).[78]

The statutes established various offices and laid out rules for training and certifying pilots, procedures for hiring professionals, and methods for research, creating a center of navigational information. As the American enterprise grew, the activities of the Casa's navigational center became more specialized.[79] First in stature and authority was the chief pilot, as a supervisor of pilots and instruments; next came the cosmographer of instruments, himself a product of the Casa's needs to improve and design instruments for navigation. Finally came the cosmographer-lecturer, who instructed pilots in the art of navigation and cosmography. The Casa's center of navigational information became the place to normalize and institutionalize knowledge and practices according to the political interest of the crown and merchants. As mentioned above, this coincided with a shift in Spanish policies in both the New and the Old World: a shift in the 1550s toward the consolidation of Spanish power in the Atlantic and in the Mediterranean.

The long-distance control of the New World was dependent upon the mapping and control of the Atlantic Ocean (see Figure 3). The political and economic needs of the Spanish empire—to extract resources from the New World—created a context in which the production of knowledge served to transform human and material resources into agents and tools of imperial domination. The mechanisms for collecting and disseminating this knowledge were institutionalized at the Casa de la Contratación over a period of fifty years. The establishment of this chamber of knowledge facilitated contacts and negotiations between the agents of the new navigational practices on the one hand and the agents of traditional practices of navigation on the other. These contacts, however, were full of tensions.

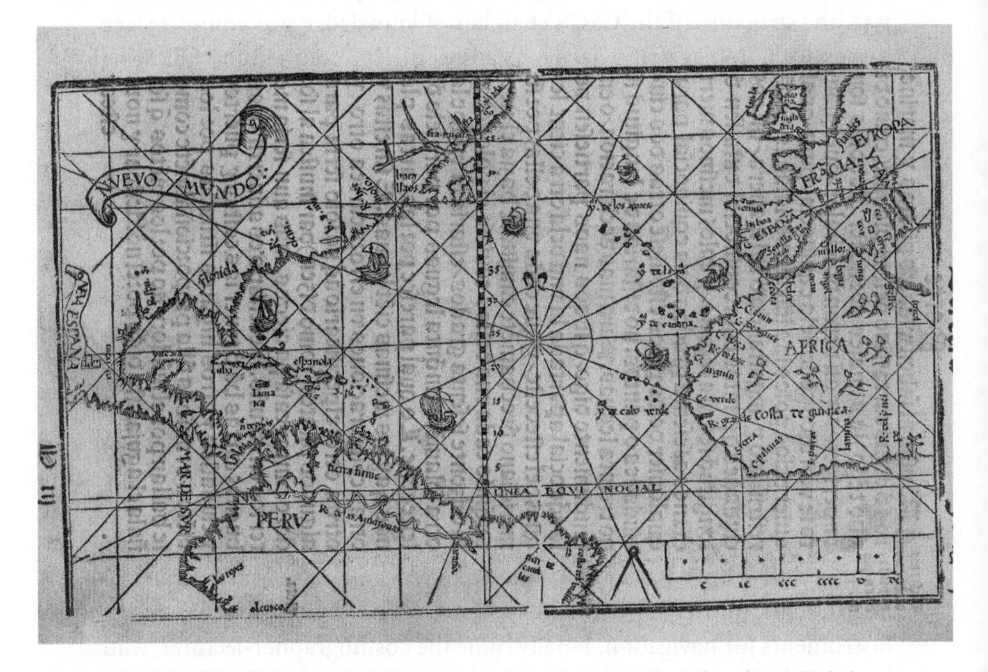

3. *The New World by 1550s in Martín Cortés,* Breve compendio de la sphera *(Toledo, 1551). Reproduced courtesy of the John Carter Brown Library at Brown University.*

NEGOTIATING NEW PRACTICES

The relationship between formal education and personal experience—in other words, the relationship between theory and practice—became a particular area of concern for all the groups engaged in the exploration and exploitation of the Indies.[80] During the sixteenth century, those with experience in the American enterprise claimed over and over again that their knowledge was more certain and truer than the knowledge of people who had never been there. In the case of pilots, the tensions were particularly intense. Pilots emphasized the role of personal experience and offered a strong counterargument to the royal officials' emphasis on theory. Some royal officials, however, did support them. In 1548 a royal official named Hernández went to the Casa de la Contratación to find a new chief pilot to replace Sebastian Cabot.[81] Hernández wrote to the king that he had been unable to find an appropriate candidate because no one fulfilled the conditions of having both experience in navigation and a good foundation in cosmography. Because he needed to suggest a candidate anyway, Hernández decided that he preferred those with experience over those with "letters."

Hernández found that the cosmographers Pedro Mejía, Alonso de Chaves, and Jerónimo de Chaves were well prepared in the art of navigation; Jerónimo de Chaves knew Greek and Latin and was a "very good philosopher and very competent in cosmography." Nevertheless, Hernández favored the appointment of Hernando Blas, who was a pilot without "letters" but "a good person" with ample experience in navigation.[82] In the end, Chaves was appointed over Blas, perhaps because he already had experience in examining pilots and making charts. Thirty years later, in 1578, the pilots (including the chief pilot, Alonso de Chaves, and Captain Juan Escalante de Mendoza) made the same argument in their answers to a navigational report by the cosmographer Alonso Alvarez de Toledo. The position of the pilots on this matter reflected an established tradition.

The center of the dispute between Alvarez and the pilots concerned which type of knowledge was most relevant for navigation. For the pilots, the chief pilot, and the masters and captains of ships, practical and navigational experience was enough. This experience, based on the pilot's knowledge of tides, currents, winds, and the ship's navigational characteristics, had been the traditional understanding of what constituted a good pilot.

The crown maintained instead that, to avoid accidents and delays detrimental to the economy of the kingdom, practical experience of pilots had to be integrated with cosmographical knowledge in transatlantic navigation. This had been the case in the circumnavigation of Africa and in voyages to the Indies: only the reading of the stars could provide pilots with accurate information about their position in high seas, far from familiar land contours.

This report shows the views of pilots and cosmographers regarding personal experience and formal training. In 1578 the crown sent Licenciado Gamboa, an official from the Council of Indies, to inspect the Casa. One of the areas of inquiry concerned "the reform of the navigation to the Indies."[83] Gamboa either received or requested a report from the cosmographer Alvarez de Toledo. Based on this report, Gamboa ordered a meeting with the "lecturer of cosmography, chief pilot, cosmographers, and pilots to discuss" Alvarez's report. This meeting produced a flurry of reports by Rodrigo Zamorano (lecturer of cosmography, appointed in 1575),[84] Alonso de Chaves (chief pilot), other pilots, and Captain Juan Escalante de Mendoza (ca. 1530–ca. 1596), who had finished a book on ships and navigation in 1575 related to the New World.[85]

Alvarez de Toledo was the cosmographer of the fleet. Some years before this report, in 1574, he had received from royal secretary Juan de Ledesma and in the presence of the royal chronicler-cosmographer Juan López de Velasco seven instruments designed by Juan de Herrera, the royal architect.[86] Herrera created a wooden circular instrument divided in 360 parts to take the longitude; an instrument to take latitudes "of places at any hour of the day"; "a small pear-wooden level"; an instrument to take "the meridian line, and the latitude, and to correct the deviation of the compass"; another instrument to take latitude "of the regions at any hour of the day"; and a metal instrument to take the meridian line and observe the deviation of the compass.[87] Alvarez's mission was to test these instruments during his voyage to Cartagena and to write a report about these instruments and the problems in the navigation to the Indies.

Alvarez identified problems in four areas of navigation to the Indies: instrument making, examination of instruments and pilots, pilots' lack of instruction, and the irresponsibility of shipmasters, who often failed to take pilots with them. The first three issues were related to the duties of the chief pilot. Lack of proper instruction in cosmography as well as the circulation of imperfect instruments forced pilots to rely on traditional methods of navigation. The Casa's poor examination of instruments and pilots, according to Alvarez's report, contributed to the situation. His inspection of instruments showed that the master portolan, the pilots' own charts, the table of solar declinations, and the cross-staff ("the instrument to measure the altitude of stars") all presented serious problems. According to Alvarez, the master chart was not regularly updated (as was required by statute 127 of the Casa); it had many errors in latitude of ports and lands. Pilots did not comply with statute 183, which ordered them to create and bring pilot books (*derroteros*) after each voyage to the Indies. The tables of solar declination were "not exact," and the cross-staff lacked precision.[88]

Yet the examination of instruments and the certification of pilots were the central issues of the report. These examinations had been intended to prevent the circulation of poorly made instruments and to certify pilots with a minimum level of cosmographical knowledge. Alvarez noted that the chief pilot failed to meet with the cosmographers to examine and certify instruments. According to statute 141 of the Casa, pilots were supposed to meet every Monday at the Casa to inspect and approve navigational instruments. Moreover, Alvarez felt that in the examination of candidates for the title of pilot, as many pilots as possible should be present rather than only six, for this number had been established "when there were not many pilots, nor were there so many voyages to the Indies."[89]

Another problem, according to Alvarez, was the illiteracy of the pilots—that is, lack of general writing and reading skills. This was the single most difficult problem for the development of the new navigational methods. Without literacy pilots would not be able to use instruments, even if those instruments were perfect. Pilots could neither read nor write their own charts. Alvarez de Toledo proposed that only literate candidates take the examinations. Yet he qualified this by proposing that illiterate pilots bring with them a literate assistant on all their voyages.[90] The main concern was the safety of the ships: illiterate pilots posed a risk. Masters also risked their ships; Alvarez claimed that some shipmasters saved a few ducados by deceiving ship inspectors and leaving Seville without hiring pilots.[91] Finally, he proposed that the chief pilot of the fleet be elected by the joint assent of the chief pilot, the cosmographers, and the regular pilots.[92] At the end of his report, Alvarez offered his services to amend the instruments and the regiment.

Alvarez's position represented the view of the crown, which supported the use of instruments and the training of pilots. His report found the problems in establishing that policy. Yet the cosmographers and pilots had a different view. Rodrigo Zamorano, Alonso de Chaves, the pilots, and captain Escalante de Mendoza responded to Alvarez's report. Zamorano seems to have worked on the report,[93] but he did not wholly agree with Alvarez's assessment of the problems. According to the reports of Zamorano, Chaves, Escalante de Mendoza, and the pilots, Alvarez was not well informed on the actual workings of the examination of pilots and instruments. Alvarez assumed that good pilots were those who had cosmographical knowledge. In contrast, Zamorano, Chaves, Escalante de Mendoza, and the pilots assumed that good pilots were, rather, those with practical experience.

Zamorano's report was much more detailed than Alvarez's, and he simply dismissed most of Alvarez's concerns. According to Zamorano, the cosmographers not only met every Monday to examine and approve navigational instruments; they also met on Fridays. The pilots had a different answer. They responded that, some time before, a few pilots used to accompany the chief pilot to examine and approve the instruments but that those pilots no longer accompanied the chief pilot because they never received remuneration.[94] Alonso de Chaves gave a similar answer, but he added that only one cosmographer was making instruments and that he inspected and approved them on Mondays and Fridays.[95] Captain Escalante de Mendoza dismissed the whole idea of inspection by arguing that pilots always looked for the best instruments.[96]

Zamorano mentioned that many pilots were summoned for the examinations and many others were allowed to attend.[97] According to the pilots, between fifteen and twenty pilots attended the examinations; Chaves essentially agreed with this estimation.[98] Captain Escalante de Mendoza felt that Alvarez was not well informed on this topic either.[99]

On the issue of illiteracy, the pilots, the chief pilot, and Escalante de Mendoza adopted a more tolerant position. Even Zamorano's report was more favorable to illiterate pilots than Alvarez's. The pilots answered that, although "it is undeniable that it is reasonable and appropriate that a pilot and master know how to read," some men made "excellent pilots" without knowing how to read; experience, rather than reading or formal knowledge of cosmography, made good pilots. Zamorano answered that "old pilots, with all their experience, even if they only read their regiments very slowly," were very good pilots. Escalante de Mendoza completely dismissed literacy; for him, good pilots were good because of their experience. Alonso de Chaves mentioned a royal decree of October 6, 1567, ordering that all candidates for the title of pilot must be literate. He thus implied that all pilots were literate to a certain degree, though some may have been slow readers (in Zamorano's characterization). Chaves recalled that he had denied the title of pilot to a certain Juan Díaz in 1573 because he did not know how to read or write.[100]

There were also conflicting interpretations of the charts. According to Zamorano, charts of the coast of Florida and the gulf of New Spain presented many "errors." The portolan of the South Sea did not include "discoveries" made from 1549 onward, including the Solomon Islands, the Philippines, and the coast of China. Zamorano proposed that all pilots coming from those regions show their pilot books. He also suggested that the royal officials should forward to the Casa copies of navigational and geographical reports sent to the king. With this information cosmographers and expert pilots would amend the master portolan on the days and times established in the statutes of the Casa de la Contratación.[101]

In contrast, Alonso de Chaves defended the extant portolan. He had been present at the time of its creation, in 1535. It was designed, according to Chaves, with the best cosmographers and "all the pilots and masters of the navigation to the Indies, even waiting for the return of those that were absent." The designers used the available "ruttiers [guides to oceanic routes], books, letters, and reports." They placed geographical features according to their latitude and longitude, their size, and their distances from each other. Chaves concluded:

> It should not be permitted that such a work [the portolan] be altered based on the opinion of one person, especially when Alonso Alvarez de Toledo has traveled only to Cartagena and back to Spain. With so little experience, he could not have verified the errors that he claims the portolan presents. This portolan was made with the agreement of over eighty men, cosmographers, pilots, and masters very experienced in the navigation to the Indies.[102]

For Chaves, knowledge about the New World was based on the increased accumulation of empirical information and its collective assessment in Seville. Eventually the portolan would have to be updated, as new information arrived; but at the time of this debate, Chaves argued, the portolan was still valid because it was the product of a community of experts, in contrast to the single report of Alvarez. Chaves's emphasis on consensus characterized the approach of the Casa.

The pilots (the voice of experience), on the contrary, acknowledged that the portolan needed some modification, because many new "discoveries" had not yet been included in it:

> It cannot be denied that the portolan lacks some trifles and that it could be mended, but these trifles are so unimportant that it does not matter whether the portolan is mended or not. For the pilots are so skillful and expert that they already navigate to the Indies without charts.[103]

Captain Escalante de Mendoza confirmed this point: "the navigation [of the Indies] is not precisely based on the portolan."[104]

Other instruments, such as the cross-staff, regiments, and tables, presented different problems. Alonso de Chaves, the pilots, and Captain Escalante de Mendoza felt that the cross-staff and regiment had minor problems. Zamorano, again, offered a more detailed view of them. The regiment presented three problems. There was an error of five minutes in the declination of the sun and thus "an error in all altitudes taken by the declination of the sun." The regiment also assumed that the sun passed through the equinox on March 10, but it actually passed on March 11. This produced an error of twenty-four minutes in the declination of the sun. Finally, the length of the day produced a difference of seven minutes every forty years in the declination of the sun close to the equinox. The cross-staff had problems with its grades. Finally, there was the assumption that the declination of the polar star was three grades and a half, but it was actually three grades and eight minutes.[105]

The pilots explained that they had been using approved instruments "and had found them dependable and without fault."[106] Escalante de Mendoza argued that the pilots always tried to use the best instruments; but if someone offered to amend the instruments, improvements should be made. With better instruments and rules for their proper use, "sailors will come to use them necessarily in their navigation."[107] The crown was always very interested in the reform of the instruments (which is why the crown sent Gamboa to the Casa in the first place). Nothing came from Gamboa's visit, however. Perhaps Captain Escalante de Mendoza was right in assuming that better instruments would create their own need: pilots would look for these improved instruments, and their own knowledge would increase.[108]

The 1578 investigation into instruments and pilots' ability to use them shows how difficult it was for the crown to impose new navigational techniques on its pilots. The difficulties, however, were not related to inefficiency on the part of the Casa officials, but rather to the particular understanding of the pilots' own navigational activities. Simply stated, they emphasized practice over theory. Against the crown's position that long-distance control of the New World was more efficient and more productive, if based on instruments and cosmography, pilots maintained the sufficiency of their own experience and their art. The crown hired and supported experts and people with direct experience in the American enterprise, either merchants living there or pilots running the ships; but the crown also wanted to discipline that experience. It was in the disciplining of experience that the crown found opposition despite its efforts.

Yet the Casa's influence on the development of navigational techniques and the collection of knowledge about the New World was successful, at least during the sixteenth century. The Casa produced knowledge of the New World in the form of books, maps, and information that circulated not only in Spain but also in Europe. It also transformed traditional navigational devices, such as charts and instruments, and created innovative practices for producing these devices. The juntas and reports made possible the master portolan and its revised versions. In the case of the training of pilots, however, the situation was more problematic. Pilots considered their personal experience and their own navigational techniques enough to accomplish their job of traveling to the New World. The Casa adapted to this kind of knowledge. The crown nevertheless created mechanisms (instruction and examinations) to incorporate the new theory of navigation into the pilots' navigational practices. Experience, however, had already become the defining element in the matrix of practices and descriptions connected to the New World (see Chapter 3).

Instruments and technology slowly became integral elements in the economic and political development of New World communities. The establishment of these communities would have been impossible without ships, charts, and compasses. The establishment of economic activities and communities depended not only on the ecological transformation of the New World and reports but also on the development of technology and instruments for exploiting natural resources, from silver to oysters and dyes.

THREE

Communities of Experts

ARTISANS AND INNOVATION IN THE NEW WORLD

In 1519 European ships arrived on Mexican shores. A man from Mictlancuauhtla who saw the ships went, of his own accord, to Motecuhzoma's palace and told him the following:

> Our lord and king, forgive my boldness. I am from Mictlancuauhtla. When I went to the shores of the great sea, there was a mountain range or small mountain floating in the midst of the water, and moving here and there without touching the shore. My lord, we have never seen the like of this, although we guard the coast and are always on watch.[1]

This New World/Old World encounter would have been impossible without "floating mountains" and the technology necessary for their use. They brought not only people, animals, plants, and viruses but also artifacts and technology from the Old World. Ships, charts, and guns—and the systems of actions to which they belonged—became the instruments of European crowns in the process of centralization for the organization and domination of distant lands in the sixteenth century. Guns and cannons provided a decisive advantage over arrows and spears. Charts helped to establish links between central authorities in the Old World and their representatives (or presumptive ones) in the New World. In the sixteenth century instruments and tools allowed their users to engage in new systems of action such as trading, colonization, and navigations.

Atlantic ships, for instance, were the products of merchant interests, shipbuilders, crown support, military needs, and Atlantic conditions.[2]

These ships opened new opportunities for European merchants and monarchs to extend their operations across the Atlantic Ocean. Portuguese merchants persuaded the Portuguese crown to support their Atlantic enterprise; the crown then appointed cosmographers, mathematicians, and other experts to devise methods to navigate the ocean safely. These officials incorporated new instruments such as the mariner's astrolabes and tables of solar declination into their new methods of navigation. The making of charts required an exchange of information between pilots, who provided geographical details about the New World, and royal officials, who organized that information into charts at royal institutions.

As they do today, instruments, tools, and technologies solidified interests and associations among certain groups of people; they fostered new visions of the world and generated new practices that in turn provided the context for even more new instruments, associations, and alliances of interests.[3] By establishing alliances and new means of exchanging information and knowledge, the people developing new instruments helped to create a practical approach to understanding nature that had social and political consequences. On the one hand, their practical approaches fostered the material benefit of private agents and, through taxation, of the royal treasury. On the other hand, it offered an alternative view to traditional religious and magical views about nature.

The instruments and technology that helped settlers in the New World were developed within a context of economic profits and political usefulness: the devices were good for the kingdom as long as they produced profits for entrepreneurs and the royal treasury. Yet these devices also emerged within the context of a legal framework of protection that made possible a system of investment in new technologies and competition among inventor-entrepreneurs. Making instruments to maximize the use of natural resources was certainly not a practice related exclusively to the American enterprise; but the American enterprise did foster it by creating the need to locate and to exploit distant resources for the empire (see Appendix 2). In contrast to humanists, whose knowledge was bound not only to texts but also to the past, inventors and developers of new technologies saw their activities in terms of future improvements.[4]

The Spanish American enterprise placed in a new political context the medieval tradition of linking practice and theory, transforming it into a set of practices that served the state, merchants, and entrepreneurs interested in innovation and profit.[5] For instance, the amalgamation process (the use of mercury for the extraction of silver) developed by a tailor from

Seville later became the object of study of physicians and naturalists such as Juan de Cárdenas (1563–ca. 1592) and José de Acosta (1539–1600). Cosmographers and pilots designed marine instruments such as bilge pumps. Mathematicians developed mills. These instruments and technology either became part of theories or were the products of theories about nature. Those who worked to improve instruments and technology advanced the idea of change and linked that concept with economic and political notions, while making technology and instruments more relevant for social actors.

Two of these actors were the Spanish crown and artisans. The crown supported new instruments and technology not only for exploiting natural resources in the New World but also for helping to consolidate the central state in Spain. The Spanish crown's interest in instruments and technological improvement had already begun at the time of the unification of the crowns of Castile and Aragon (1479) and was in full force by the end of the Reconquista (1492) and the beginning of the American enterprise (1493).[6]

Isabella and Ferdinand fostered the construction of roads, bridges, canals, aqueducts, and mills as a way to achieve the political and economic unification of the Spanish kingdoms. The rate of bridge and road construction in Spain then declined during the reign of Charles V, because his priority was the defense of the empire against the Ottoman empire and the Protestants. Charles V's assumption of the throne, however, made possible relations between Spain and artisans from Flanders, France, Germany, and Italy—despite royal attempts to limit the work of foreigners in Spain and the New World. Recall the Welser bankers sending French experts in the cultivation of pastel to New Spain; at some point they also sent fifty German mining-masters to America.[7]

Artisans gained in status and autonomy as the Spanish crown requested their services. In Aragon a group of experts in the mechanical arts—Jerónimo Girava, Pedro Juan de Lastanosa, and Pedro de Esquivel—solved practical problems in architecture and engineering based on Italian and Dutch models. They were summoned first to the court of Charles V and later to the court of Philip II.[8] Philip ordered Esquivel and Lastanosa to work on the reorganization of the Spanish transportation system. Although the lack of roads in some areas and the poor condition of the existing roads halted Esquivel and Lastanosa's project, they were part of the group of technological experts working at the court of Philip II.[9] Esquivel attempted to survey the towns and cities of

Spain.[10] Through these activities, the crown sought to expand its control over Spain's human and natural resources.

The crown protected the work of artisans who developed efficient and profitable technologies and instruments. As in the case of Florence and Venice, the Spanish crown granted royal licenses to inventors as a form of protection similar to a patent. The royal licenses had a standard legal structure: they granted the inventor the right to use an invention in a certain place during a limited time and instituted fines and punishments for those who tried to use or copy a similar device without authorization from the inventor. Thus, the royal license provided the inventor with the legal means to take anyone who copied and used his device to court.[11]

In general, the information from the royal licenses was very succinct. Devices are described only briefly, perhaps to protect them further. Yet there are few cases that allow us to look more deeply into the sixteenth-century culture of inventions. In many instances, the new devices licensed were only improvements over already-familiar and common versions of the same thing. Most of the instruments belonged to one of these four groups: diving instruments to fish oysters (pearls) or instruments to rescue objects from the ocean (Appendix 2, Table 4); devices to improve mills (Appendix 2, Table 5); navigational instruments (Appendix 2, Table 6); and instruments to exploit metals (Appendix 2, Table 7). Other examples outside of these categories include potentially useful inventions (Appendix 2, Table 8), such as a lantern or a seawater purifier.[12]

Although the scholar Francisco José González, following Zilsel, agrees with the general characterization that I present here, he dismisses the Spanish activity too quickly, as those who follow Merton usually do. "Spain," González writes, "became the center of the Counter-Reformation and thus, in its opposition to the Protestant Reformation, stopped any attempt to renovate scientific activities [in Spain]."[13] The emergence of empirical and mathematical scientific activities was not the result of Protestantism (nor was the lack of them the result of the Reformed Catholicism). Rather, the emergence of empirical scientific activities, in particular, resulted from the commercial and imperial activities of the sixteenth and seventeenth centuries. The early Scientific—that is, empirical—Revolution that took place in Spain became a key element in later developments, as the Spanish scientific books translated into Italian, French, and English (among other languages) testify. Spain did revise scientific practices; and what happened to those practices during the seventeenth century (lack of practitioners, institutions, and resources) is part of a history that has more to do with economics and politics than with religion (and religion was, as it is today, politics).

MAKING ARTIFACTS

Early in the sixteenth century, explorers began to use tools for increasing their profits in the New World. From at least the early sixteenth century, fishing rakes were used in the Caribbean and on the coast of Andalucía. In 1509, for instance, the explorer Alonso de Ojeda used a rake to fish oysters at Araya Point.[14] Yet it was only in 1520 that the pilot and explorer Juan de Cárdenas (d. 1540) obtained the first license to use a device, probably a rake, to fish oysters in the Caribbean. This was not a separate license but was included within the royal decree granting Cárdenas rights to trade in the New World (Appendix 2, Table 4).[15] The practice of granting separate licenses for using new instruments appeared in the Atlantic world only in the mid-1520s as a result of private requests for protection.

According to Cárdenas, oysters near Araya Point and Coche Island (near Venezuela) were too deep for divers (indigenous slaves). Cárdenas claimed "to have a certain type of device" useful to collect these mollusks, and he asked the king for a license to use it. The crown granted the license, because Cárdenas claimed that his instrument could fish oysters in areas inaccessible to divers and increase the royal treasury through "the fifth that corresponds to us"—that is, one-fifth of the profits from the pearls.[16]

Cárdenas's license not only linked instruments, commodities, and profits with the crown but also established a tie between the expert in the New World (Cárdenas) and the royal bureaucracy in Spain, the same kind established years later between royal officials and Antonio de Villasante, Gonzalo Fernández de Oviedo, and the pilots of the Casa. The crown supported the Cárdenas device because it appeared profitable, but his license did not protect him from losing profits to those who might copy his instrument. It only granted him the use of the device and established his obligation to pay royal taxes.

The first royal license that actually protected an instrument for use in the New World appeared in 1524, the same year in which the crown formally established the Council of Indies and the Inquisition summoned a meeting at Valladolid to decide about the orthodoxy of the writings of Desiderius Erasmus. With his critiques not only of the church but also of wars of conquest, Erasmus was a significant influence at this time in Spain; the meeting at Valladolid proved as much. Personalities such as Alonso Manrique the inquisitor and Archbishop Fonseca of Toledo, among other influential court officials, supported Erasmus and his ideas in Spain.[17] In this context of debate and reform, the crown was also expanding its own imperial activities by bringing artisans into the American enterprise. Patents for the protection of inventions and technologies are one example of this expansion. The royal license of 1524 mentioned above can be considered the model for subsequent licensing in the New World.

Such licenses incorporated elements from the inventor's proposal—the expansion in crown activities in the New World came from below. In 1524 the cosmographer Diego Ribeiro offered to make "metal [bilge] pumps to drain ships." He claimed that each one of his metal bilge pumps would "pump as much water as ten of the wooden pumps that are in use now" and with only "a third of the people" needed to operate traditional pumps. The bilge pumps of the Portuguese cosmographer Simon Fernandez might have inspired Ribeiro.[18] In any case, Ribeiro claimed that his pump was his invention and that it would be lighter, easier to handle, and more efficient than traditional wooden pumps. He offered to "test two pumps at" his own "expense" and asked the crown that, if the test "turns out true" (*la prueba . . . saliendo verdadera*), he be granted a license for making the pumps.[19]

In 1524 Ribeiro received a conditional license (confirmed in 1526).[20] The conditional license explained that, taking into consideration Ribeiro's service to the crown

> and the benefit and universal good expected from . . . [the pumps] to come to us, our subjects, and residents of our kingdoms, we say that once you make those pumps as you bound yourself to do, and conduct a trial with them in Coruña or Seville, where they would be examined and seen by the people we would designate for the purpose of that experience, and if they approve and consider them useful for navigation and if we want to use them, we would grant you . . . sixty thousand maravedís.[21]

In 1531, when Francisco Pizarro landed on the western coast of the Inca empire, Ribeiro had his first pumps ready; the crown ordered the Casa to investigate and to name appropriate people to test the pumps.[22] A few months later, the crown once more gave orders to the officials of the Casa:

> To appoint shipmasters, sailors, and people with knowledge and experience in navigation to test and experience the aforementioned pumps in a vessel.[23]

The crown also ordered the Casa to send reports of these trials to the Council of Indies. Recall Villasante's case: the crown ordered Villasante to send samples of balsam to physicians and hospitals in Castile; these physicians in turn were asked to write reports about their tests. A similar situation took place with Ribeiro and his pumps. In both cases the crown brought together experts, merchants, and artisans to examine artifacts and commodities. An empirical culture emerged around artisans and royal officials tied to the Atlantic world for the study of nature, with institutions supported by the crown.

The pump's first trial took place on November 25, 1531, in the ship *Santa María del Espinar.* Three royal officials from the Casa and five experts in navigational matters tested the pump and found it to be more efficient and easier to handle than wooden pumps.[24] However, the council was not convinced by the results. In 1532 it ordered the Casa to pay Ribeiro sixty thousand maravedís as a preliminary payment; meanwhile, the council would organize more extensive tests.

On May 4, 1532, Ribeiro delivered a bilge pump of 303 pounds to the captain of the ship *Alta Mar.* The vessel headed for Santo Domingo but turned back to Seville before it could finish its voyage because it was taking on water. According to the reports from the captain, three pilots, and five sailors, the pump's great efficiency was the reason the ship made it back to Spain. The Casa sent the *Alta Mar*'s report to the council with its own positive opinion about Ribeiro's pumps; but, alas, Ribeiro never enjoyed the benefits of his successful invention: he died on August 16, 1533. Ribeiro's heirs and legal representative Diego de Oliver continued Ribeiro's process before the crown.[25]

The tests conducted by experts and royal officials on Ribeiro's bilge pumps constituted an example of a practice later institutionalized at the Casa for validating new instruments and technologies. The crown expanded empirical practices already in use among artisans and incorporated these testing practices into the state's regulation of instruments, technology, and inventors. The establishment of testing rules had been in part the result of Ribeiro's own suggestion to conduct tests. These rules compelled artisans to take their instruments before royal officials, where experts would test their actual performance. Inconclusive tests led to more tests under better conditions. Test results provided royal officials with the information necessary to validate the claims made by the inventor. By the 1570s the Council of Indies already had a "rule that those who have [inventions] must appear before the council for examination."[26]

In 1535 Vicente Barreros, a carpenter, offered cheaper and better wooden pumps than those of Diego Ribeiro. Barreros asked for a license similar to the one granted to Ribeiro. The crown had created a legal precedent for this request. In 1526 the crown had stipulated that if someone "wanted to compete and challenge" Ribeiro's claims about his pumps, Ribeiro "would allow him to do it"; and if "within two years someone came up with a better invention, he would have" the same prerogatives granted to Ribeiro.[27] The crown, following the established procedure, ordered the Casa to investigate whether Barreros's pumps were "more useful, better, and cheaper" than the metal pumps designed by Ribeiro.[28]

The Casa conducted tests to verify Barreros's claims about his pumps. After these tests and a positive recommendation from the Casa officials, the council granted Barreros a license to make and sell his pumps for five years.[29] The crown had managed to promote competition between inventors for improving the efficiency of bilge pumps. Royal licenses set the stage for inventing and improving instruments and technology and established empirical procedures to validate them. These procedures emerged first from the interrelations among artisans, merchants, and royal officials engaged in different activities: merchants searching for medicinal plants, pilots making charts, carpenters devising bilge pumps. Then the same empirical procedures became common in making charts, studying medicinal herbs, and investigating the fauna, flora, and geographical characteristics of the New World. The crown, together with merchants, artisans, and royal officials, sought to transform personal experience (in the form of reports or claims about instruments) into formal knowledge. The procedures at the Casa's chamber of knowledge disciplined personal experience within a complex social setting of experts and officials.[30]

The crown granted a license once the instrument had been tested and its efficiency had been determined by a group of experts. This process made experience the criterion for determining the validity of information, knowledge, instruments, and technology. Furthermore, the proven efficiency of new technologies and instruments became the criterion for validating them. To test for efficiency, the crown established a system of verification based on trials and supported individuals who attempted to make instruments for a more efficient exploitation of nature.

At the same time that Ribeiro received his license (1526), the crown granted the Sevillian Juan Fernández de Castro a license to use an *ingenio* or *artificio* (instrument or device) for extracting gold from rivers and ponds.[31] Castro claimed that the gold of "rivers, ponds, and creeks" from the Caribbean islands was not exploited at all, for lack of "industrious and ingenious" people (surely referring, incidentally, to his fellow Spaniards, because the indigenous people had been exploiting gold for years).[32] The crown granted him a license provided that this was in fact "a new instrument, not invented before in these parts [Castilian kingdoms], nor in the Indies." He had "to show the said instrument [*ingenio*] or device [*artificio*] before us and before [the members of] our Council of Indies or before our officials . . . of the Casa de la Contratación within four months." Eight months after that, he could begin using it in the New World.[33]

The king showed great interest in Castro's proposal. If the instrument proved to be as efficient as Castro claimed, it could mean an increase in revenues for the royal treasury. Castro was the intellectual owner of his instrument, but a provision in his license reserved the possibility of the crown's granting the use of this instrument to the inhabitants of the Caribbean islands:

> If, [after] the fruits that come from the said instrument are seen by us, we give general or particular license to the residents and inhabitants of the . . . [Caribbean] islands and any other person for using . . . [the instrument], all people who use it would pay you and your heirs four percent of all that they obtain with said instrument.

Castro's and Ribeiro's licenses followed a similar pattern. Both licenses linked instruments, innovation, and profits, as Cárdenas's 1520s license did. Castro's license compelled the inventor to "show" his device to royal officials, something that Ribeiro had suggested that he would do himself. Castro's license, finally, protected his instrument by establishing his exclusive right of use. The crown hoped to make the instrument available to other people, however, and thus increase royal revenues as well as individual profits.

By supporting and protecting instruments and their makers, the Spanish bureaucracy fostered a practical approach to studying nature. Entrepreneurs and bureaucrats approached nature as a repository of useful resources, the source of material benefit (to the crown as well as to entrepreneurs). "Industrious and ingenious people" such as Castro, who could create *ingenios* or *artificios* "not seen until now," received the protection of the crown. As seen in the case of Barreros, the crown also legitimized innovation as an efficient means of improving instruments and technology. Making instruments and developing technology fostered the notion that practice and theory could be accumulative and subject to improvements. In 1551 the cosmographer Martín Cortés claimed that during Charles V's reign "Spain looks renewed, and in all the mechanical arts [Spain] has improved and is better."[34] Such was the case with mining technologies developed in the 1540s and 1550s.

COMMUNITIES OF EXPERTS

As Ribeiro, Barreros, and Castro developed their instruments, Villasante performed his tests with balsam, and the pilots at the Casa made charts, the Spaniards found mines in the New World and began to extract minerals with improved technology. Juan Tetzel in Cuba and Bartolomé de Medina in Mexico developed new methods for the extraction of metals; and, in doing so, they (or, more specifically, their methods) helped to establish communities of miners, merchants, and royal officials with similar goals and practices regarding the natural world.[35]

Copper was discovered in Cuba in the late 1520s and exploited beginning in 1534. These mines became the New World's most important copper source until the discovery of copper in South America.[36] Around 1542 Juan Tetzel became a major player in the fledgling Cuban mining industry. He found the mines "moderately rich," but their exploitation was difficult. Tetzel "made preliminary experiments [*experiencias de prueba*] with this metal . . . to discover its nature and quality," without success. According to Tetzel, there were neither technical resources nor "minds for the understanding" of this matter in Cuba. Between 1545 and 1546, just before the war between Charles V and the Protestants (1547), Tetzel went back to Europe with samples of copper to "test and try" the metal. He "travel[ed] around the land of Germany, where there are miner masters, experimenting with the metal." Finally, "having experimented with the metal and [having] determined the nature of it," Tetzel "learned the secret of it after much work and personal and patrimonial expenses."[37]

Tetzel approached the Spanish crown and offered his knowledge in the development of the mines. He wanted to take to Cuba and cover the expenses of masters and "all types of officials" and tools for exploiting the copper mines. The crown, in return, granted Tetzel numerous incentives: the rights to ten mines; all the wood he needed; freedom of movement in the Caribbean and between the New and Old Worlds for himself and his workers; the right to build mills and residential buildings around the mines; exemption from taxes (*almojarifazgo*) on tools and clothes for four years and on copper sent to Spain for the next ten years; and half the royalties from any copper produced by those who would learn his method of extraction and production.[38] Tetzel was given only one year to leave Spain and two years to begin production, and he was required to be back in Cuba by early 1547.

During the first few years, Tetzel "exploited great amounts of copper and continued" the work of the mines "until French corsairs plundered and robbed the city of Santiago." In this raid Tetzel lost most of his melted copper and estate. Later, hurricanes destroyed "two times the mills [*ingenios*] that he had for exploiting said mines." Finally, all but two "masters and officials that . . . [Tetzel] had brought" to work in the mines died.

Needless to say, Tetzel had run out of resources for working the mines—yet by June 1550 he had not disclosed "the secret method appropriate for melting and for easily working [the ore]" to other residents of the city of Santiago. In a meeting at the town hall, residents "claimed and claim that Juan Teçel had the obligation to disclose the secret of the condition and the smelting process for the aforementioned copper, so that they could enjoy the exploitation of said mines," as was the "intention of the king."[39] This was not an unusual request, for such provisions were standard features of licenses. Allowing other entrepreneurs to use new instruments and technology was an essential component of economic development.

The residents of Santiago offered Tetzel a contract. He would first have to "teach the method and secret to melt and work the copper," so that "everybody who wants to try this [the exploitation of the ore] can work" the mines. Second, he would have a year and a half to teach his method "to the black slaves" that "the residents of the city and island wanted to give him" for that purpose. Tetzel would pay thirty pesos of gold for each of these slaves. Whatever the slaves produced in the mines would belong to him. In exchange for Tetzel's teachings and supervision, the residents would give him and his heirs, "perpetually," 3 percent of the melted copper.[40]

In 1569, Tetzel returned to Madrid to request a confirmation of his 1546 capitulation, which was the basis for his contract with the town hall of Santiago. He also requested one hundred licenses to buy slaves free of taxes as well as other financial incentives.[41] The king confirmed his capitulation of 1546 but did not grant the other requests.[42] Tetzel died, at an unknown location but possibly in Madrid, between March 1571 and January 1572.[43]

As the case of Tetzel suggests, the act of invention took place within a context of multiple interests defined by entrepreneurs, experts, and royal officials. The interactions among these groups created a situation in which technological innovation occurred: entrepreneurs were interested in exploiting natural resources; royal officials were interested in increasing the royal treasury and in establishing the legal means for controlling the use and application of new instruments and methods; and artisans were interested in finding ways to create useful technology, secure property rights over it, and make a profit. The common ground for these actors was a commodified understanding of nature and an empirical and practical approach to its exploitation.

One of the differences between Tetzel's case and those of instrument inventors is the significant place of personal experience and contacts with foreign experts in Tetzel's account. Ribeiro and others developed instruments based on their own mechanical knowledge of mills and pumps. Lastanosa, for example, claimed to have tested his mill before requesting a license, and Ribeiro offered to conduct tests to obtain his license. Tetzel, however, used not only his own experience but also tests performed by experts of the Holy Roman Empire: empiricism based on collaboration was his method for obtaining a royal license, for convincing royal officials that his method was useful and valid.

During the sixteenth century, foreign experts in architecture, alchemy, medicine, gardening, and navigation came to work in Spain and in the New World. Italian and Portuguese experts such as Amerigo Vespucci and Diego Ribeiro worked at the Casa de la Contratación; Charles V granted a license to the Welser bankers to send fifty German experts to the New World. Thus, Juan Enchel, Juan Tetzel, and Miguel Redelic worked in New World mines from the 1530s onward.[44] As mentioned previously, French experts traveled to New Spain in the late 1530s to initiate the cultivation of pastel there. In the late 1540s Maximilian of Austria brought German experts to work in Spain, during the regency of Maximilian and his wife, Maria (daughter of Charles V). Before 1555 at least fifty-seven Flemish passengers, twelve Germans, and one Alsatian went to the New World.[45] Philip II sent Spanish experts abroad to collect information about gardens, and he attracted Dutch and Italian gardeners to work at his own gardens.[46] In sum, despite Castilian royal restrictions on foreigners in its oversea kingdoms, the crown maintained a margin of flexibility for the movement of foreigners into its territories.

Tetzel's activities in the 1530s and 1540s belonged to this cultural context. He brought his expertise from the Holy Roman Empire to Santo Domingo. When he could not solve a technical problem, he sought help from other experts in Spain and in the Holy Roman Empire. Once he was able to solve his problem, he went back to Santo Domingo and profited from his knowledge—first alone, then later by selling his knowledge to Santo Domingo's residents through the training of African slaves. Personal experience and testing became the common element for validating knowledge within Tetzel's circuit of contacts. This loose and informal community—a republic of experts, whose contacts and methods the crown supported—shared not only a collaborative and empirical approach to nature but also a positive understanding of change and of the uses of technology in society. This is especially apparent in the case of the development of an amalgamation method by a Sevillian tailor through craft traditions.

INNOVATION AND TECHNOLOGIES

In the early 1540s, when Tetzel first started testing samples of copper in Germany, German and Spanish miners began testing new methods for exploiting silver in New Spain. Silver had been discovered there in the mid-1520s, yet the mines were only small enterprises. The discovery of silver in Taxco in 1534 marked the beginning of New Spain's large mining activity.[47] Years later, Velázquez de Salazar, in his report to Juan de Ovando (president of the Council of Indies, 1571–1575), explained that the production of silver entered a period of crisis in the early 1540s when working the exhausted mines became expensive and unproductive. Around 1542 Juan Alemán, after receiving a "report sent to him from Germany," proposed a new method (based on bone ash and lead) to mine poor silver ore.[48] On June 8, 1550, the German miner Gaspar Loman obtained a license to exploit silver for six years using Juan Alemán's method.[49] The method worked well for a few years, but then the production of silver declined again. Fortunately for royal coffers, the culture of invention created incentives for artisans and professionals to develop new instruments and technology. In 1553 the tailor Bartolomé de Medina (ca. 1497/1504–1585) arrived in New Spain with the purpose of developing a cheaper and more efficient method of mining New Spain's silver.[50]

Medina traveled with his own means, as Tetzel had traveled, to the New World; but he expected to find official protection and thus economic advantages once he had found a more efficient method for exploiting silver.[51] In the early 1550s a scribe at the court of viceroy Don Luis de Velasco recorded the story of Bartolomé de Medina:

> Being in Spain, he had news of the method employed to exploit gold and silver [in New Spain], and of the great cost . . . of it; and to know if this was so, he came to New Spain to see it with his eyes and to endeavor how to exploit those metals at a lower cost, and thus with great industry and care and work of his person and expense of his patrimony he understood, on account of the experiences mentioned, a method to exploit those metals by using quicksilver and obtain from them all their ore . . . with less cost in people and horses.[52]

After great work and costly experiments, Medina developed a method using mercury for smelting silver, which is called the amalgamation method. In New Spain, this method became known as the patio system, because it was used in small yards or patios. It consists first of combining silver and mercury—since mercury absorbs silver—and later of separating them.

Perhaps Medina, as a tailor, already had experience in smelting small quantities of silver and gold to use in making dresses.[53] He certainly had the help of a German expert known as Maestro Lorenzo, with whom he probably discussed the possibility of using mercury.[54] By late 1554 or early 1555 Medina had developed the amalgamation process in New Spain. His method reached Spain in 1557 and Peru in 1572 but did not reach Central Europe until the 1780s—two and a half centuries later.[55]

Medina's method consisted of preparing a mix of silver ore, water, large amounts of salt, and mercury. This mix was prepared in small boxes or enclosures on an outdoor patio that was divided into shallow rectangular pools. The process took about twenty-two days. This system prompted the construction of a system of patios surrounded by administrative and manufacturing buildings.[56]

Medina's innovation consisted of the adaptation of the amalgamation process already known to alchemists and German miners and discussed in the work of Vannoccio Biringucci (1480–ca. 1539), published in 1540, for the industrial production of silver.[57] Medina's method was implemented quite successfully: by the late 1550s more than 120 people were already using it; by 1562 in Zacatecas alone there were thirty-five schemes based on the amalgamation process.[58] Medina's technology transformed social relations in New Spain and Peru by making the exploitation of silver viable again as well as accessible to new entrepreneurs. This is perhaps the first significant industrial innovation of the early Scientific Revolution.[59]

Medina's system was exported to Peru in 1572. There the system not only induced an explosion of improvements and royal licenses but, more importantly, transformed social relations in the viceroyalty. The production of silver in Peru had fallen from 379,244 marks in 1550 to 114,878 marks in 1572. After 1572, however, production rose again (to a peak of 887,448 marks in 1592).[60] Indigenous labor was organized into a draft-labor system (called *mita* in Peru and *repartimiento* in New Spain) to meet the new levels of production. The *mita* (Quechua for "turn") was "a draft labor scheme designed to bring to Potosí some 13,400 male Indian workers annually."[61]

In 1569 Cardinal Diego de Espinosa (president of the Council of Castile), Francisco de Toledo (viceroy of Peru), and Juan de Ovando (soon to be appointed president of the Council of Indies) discussed, among other American policies, the implementation of the new refining (process) and the possibility of intensifying the draft labor system in Peru.[62] Viceroy Toledo subsequently received instructions from Philip II to implement both systems in Peru, which he did in late 1572.[63] The area designated for supplying the labor force extended 800 miles from Cuzco in the north to Tarija in the south and about 250 miles across the width of the Andes. Workers (males between eighteen and fifty) had to go to Potosí for a year; this provided around 13,500 men a year working in groups by turn, one week on and two off.[64] Their wages were set daily, according to the job. In the late 1570s a great influx of silver began entering Spain, which would help finance the Armada of 1588 against England. The basis of this economic development was the empirical method for studying nature established in Spain by the mid-sixteenth century.

Medina's method not only transformed social relations in the viceroyalties; it generated competition for developing more efficient and cheaper ways to implement the amalgamation process. Once the patio system was established in New Spain, a series of other inventors sought to improve it, mostly by providing more efficient and cost-effective uses of mercury (Appendix 2, Table 7). In 1559 or 1560, for example, Alfonso Martínez de Leiva requested a license to carry out some experiments in the Zacatecas mines to reduce the amount of time the amalgamation process took. The viceroy granted him a license to carry out experiments for three months and, if these were successful, exclusive rights to his method for six years.[65] Other miners sought ways to reduce the use of quicksilver (mercury) and thus to reduce the cost of the process. In 1559 Pedro González and Alonso de León began using sieves to reduce the ore to a fine powder, thus making the reaction with mercury more efficient.[66] In 1562 Pedro Díaz de Baesa used fluted earthen jars to save quicksilver in the last step of the amalgamation process. On July 10, 1563, Juan de Plascencia licensed a new sieve for metals and an instrument for spinning the mesh used in his improved sieve. In 1567 Leonardo Fragoso and Cristóbal García developed a new metal washer for reducing the loss of quicksilver. Gaspar Herrera developed a method for reducing the amount of quicksilver in 1566.

Some of these improvements required great work and expertise; in the context of competition generated by the royal patents, people carried out their experiments and work in secret and at great personal expense. The viceroy Martín Enríquez commented that a certain Juan Capellín worked for "nine years in many inventions and buildings in his house and some secret places to learn and understand" the method for reducing the time of the amalgamation process and thus the amount of quicksilver used in it. He designed a mill with metal mallets and an instrument to retrieve quicksilver used during the amalgamation process, and he obtained two licenses in 1576.[67] Bernardino Santacruz developed a "wooden box" and a "small tank" with a "secret" to save quicksilver. The viceroy found his invention useful and granted him a license on September 3, 1580.[68]

Miners also sought to reduce the use of human labor in the process. Alonso de Espinosa developed a system of rakes pushed by beasts of burden to mix the ore with quicksilver. He obtained his license on February 22, 1561.[69] In 1563 Juan de San Pedro designed a method for accelerating the amalgamation process based on the use of magistrals (roasted and pulverized copper pyrites).[70] Juan de Plascencia, who worked with Medina in the Pachuco mines, developed horse-powered bellows for mine furnaces, "bringing great utility to the natives [of the land] as well as to the king and the Spaniards."[71] Still others worked to improve methods for obtaining resources (such as salt) necessary for the amalgamation process. Marcos de Ayala, Martín Alonso, and Pedro de Ledesma licensed a method to obtain pure salt in 1564.[72]

The crown fostered the development of empirical practices and rational calculation regarding nature by linking profits and the material benefit of the individual to instruments. The crown also supported the expansion of empirical practices by opening the culture of inventions to those who successfully tested instruments and by protecting legal rights over instruments and technologies. Simultaneously, the use of instruments and technology promoted particular views toward nature. On the one hand, nature was a collection of resources ready for exploitation. Minerals, for instance, were intensely exploited, and mining technology was increasingly improved to extract precious ore such as silver from previously unworkable mines and metal residues. On the other hand, nature was also a collection of limited resources needing administration. Instruments could damage natural resources. Fishing rakes, for instance, caught not only oysters but also everything else on the ocean bed, destroying the very environment in which new oysters would grow.

Despite the crown's and entrepreneurs' views of instruments as fostering royal profits and the material benefit of individuals, each new instrument or technology was implemented in a particular locality and subjected to local interests and practices. Local usage of instruments and technology promoted the practice of administering nature as a counterpart to the notion of instrumental exploitation promoted by the central state.

ADMINISTERING NATURE

In addition to destroying the natural environment of oysters, fishing rakes also challenged the traditional use of African slaves and indigenous people as divers. This was true for the instrumental manipulation of nature more generally, especially when the central government and peninsular entrepreneurs promoted the use of instruments for increasing profits and for the political good of the crown. The crown could not, however, impose the use of an instrument if the instrument endangered the way of life of an area's residents. Local entrepreneurs had to balance their own desire for profit against both natural sustainability and local social organization.

Despite Caribbean residents' opposition to the use of fishing rakes, the crown continued to grant licenses for them. The case of Domingo Bartolomé, "Frenchman," and his device for fishing oysters (called the *tartana*) is a good example of the crown's promotion of instruments.[73] In 1592 the crown granted Bartolomé a license subject to successful testing and ordered the royal official Simón de Bolívar—on his way to Venezuela—to stop on Margarita Island to perform the testing. Bolívar's mission was to test the instrument

> in those places where the fishing of oysters is more scarce as well as in the other [fisheries] by their degrees [of depth] to the deepest fishery. [This test should be carried out in such manner] that no port or place be left without a test to understand clearly how beneficial this [instrument] would be and if it would be useful, as its inventor claims.[74]

Furthermore, the scribe of the city council accompanied Bolívar and Bartolomé in order to provide a detailed account of the tests, including information such as distances to the sea-floor, durability, and costs. The crown commanded Bolívar and Bartolomé to discuss the results of their tests with the governor of Margarita Island and the royal officials of the treasury and to determine with them the usefulness and costs of the instrument. If the experiments showed that the oyster rake reduced the time and cost of fishing for pearl oysters, the crown could authorize its use.[75]

The tests began on November 10, 1593. Several royal officials, residents, and a friar came to the port of San Pedro Mártir to see the experiment. At least two local experts, Domingo Marcano and Melchior López, accompanied Bartolomé and Bolívar in a canoe to test the instrument. They first tested the *tartana* at a depth of seven fathoms, at a place rich in oysters. Once the fishing rake was submerged, the canoe navigated "two or three shoots of a crossbow"; then Bartolomé retrieved it and found it filled "with many oysters." Yet when they opened the oysters, they found few pearls; Marcano and López explained that these were "new and growing" oysters.

The party tried the instrument again, this time at twelve fathoms, a depth inaccessible to divers. On this occasion, the rake pulled in few oysters and some "sea-creatures of different kinds." In a third attempt, at fourteen and a half fathoms, the instrument removed "things" from the sea-floor but no oysters. Marcano and López thought that the lack of oysters in the rake was not the fault of the instrument but rather a sign that there were no oysters at that depth. At this point, Bolívar decided to stop the tests and try again a few days later.

A new set of tests began on December 1. Again, many royal officials and residents were present during the tests. In the first two tests, the rake caught "starfish that are big and heavy crustaceans," crabs, and "things" from the sea but no oysters. They tested the *tartana* about five times at eleven and a half and twelve fathoms without success, precisely the depth at which Bartolomé claimed his instrument would be very useful.

A week later, Bolívar ordered fourteen officials and residents to meet at the main church of the town of Asunción

> to speak and declare their opinions [about "the instrument and net to fish pearls"] and if they want to use it and if it would be useful and appropriate for fishing oysters.[76]

The town residents, after having seen Bolívar's royal decrees and discussing them among themselves, reached a consensus. They considered that Bartolomé's "device was only a web or hammock of little profit." Quite bluntly, they considered it useless. In the first test, it fished "only growing oysters and thus could deplete the oyster beds in two years." In the ensuing tests, it did not fish oysters at all. The device showed that there are no oysters at more than nine fathoms deep. And at that depth slaves could still swim. The residents told Bolívar that the "device" that they needed was to bring more slaves.

Despite the lack of interest in his device, Domingo Bartolomé insisted on further tests. Bolívar, following royal instructions, granted Bartolomé another opportunity. Jhoan Rolon and Nicolao Reinaldo "of the Savoyan nation," both "experts and with experience" in the coasts and oyster beds, were ordered to accompany Bartolomé, as witnesses to his tests.

The results of these tests were also a failure. The instrument either did not fish oysters or fished "new and growing oysters." A couple of times it was entangled with rocks in the sea-floor. After these tests, Bolívar commanded that Bartolomé's instrument be burned. At the crown's behest, Bolívar ordered Bartolomé back to Spain on the first ship available.[77]

As the case of the oyster rake demonstrates, not all new inventions proved to be successful—but even the unsuccessful ones provided the crown and its subjects with new information about nature. The inhabitants of Margarita Island learned that oysters do not live beyond a depth of nine fathoms and that the *tartana* could deplete the sea of growing oysters. The residents were not simply consumers of natural resources but also administrators of nature. In the process, they (and residents of other similar communities) learned more about nature and were able to define their own positions better. Administering nature implied learning the limits and workings of nature and how to use instruments appropriately.

SCHOLARS AND ARTISANS

The practices described in this chapter influenced not only the behavior of those directly engaged in the American enterprise but also the work of natural historians. In particular, natural historians such as Juan de Cárdenas and José de Acosta, for example, became interested in explaining the amalgamation process. Natural historians became brokers between the culture of inventors, with its tests and informal circuits of information, and the textual culture of natural history.

Juan de Cárdenas (1563–ca. 1591) was a significant intermediary between miners and scholars. Cárdenas went to New Spain in 1577. He studied medicine at the University of Mexico (established in 1553). In 1589, at age twenty-six, he finished his book *Problemas y secretos maravillosos de las Indias* (Problems and Marvelous Secrets of the Indies), which the Mexican printer Pedro de Orcharte published in 1591.[78] Cárdenas divided his text into three books: in the first and third, he discussed the characteristics of New Spain's land and people, respectively. In the second book, Cárdenas discussed "copiously the exploitation of metals, giving the reason salt is cast in the piles of ore for obtaining silver, and why so much quicksilver is lost" during the amalgamation process.[79] Many miners had been working on these problems, either by developing different methods to reduce the use of mercury or by developing methods to obtain pure salt. Cárdenas explained these problems, using a theory shaped by practical developments in the mining industry.

First, Cárdenas discussed the uses of salt in the amalgamation process. "What is under consideration . . . [now] is to know what mystery or strange secret exists between silver and salt, that it be necessary to throw salt to separate [the silver] from the ore." He framed the problem in terms of a secret relationship between salt and silver, according to the classical theory of the elements and their qualities and perhaps with some alchemical influences. Cárdenas continues:

> To say that between salt and silver exists some sort of friendship and harmony, and that by means of this silver is extracted, does not lead anywhere, because silver is cold and wet and salt is hot and dry; . . . [salt] is of the nature of water, the other of water and earth. Thus the effects of one and the other are very opposite. According to this: what likeness can be between the two for their mixing?[80]

Cárdenas wanted to explain the role of salt in the amalgamation process. He ruled out the possibility of salt mixing with silver because salt and silver present opposite qualities. For Cárdenas this explanation "does not lead anywhere."

To figure out the role of salt, Cárdenas explained the process:

> [Miners] take the ore and grind it very well; then they mix it with salt, and with some pounds of quicksilver according to the ore's quality. After a few days, it is assumed that the ore is ready (that is, when silver and mercury are embraced), and then the miner has the ore washed with water. The mud and sand leave with the water, and in the bottom of the tank remains a cluster or *pella* of silver and mercury. Finally the miners separate the silver and mercury with fire.[81]

Cárdenas further explained that silver and mercury had between them "friendship, analogy, and fitness" on account of their great similarity, and this is why they "embrace" each other during the amalgamation process.[82] Mercury and silver attract each other because they have similar properties. Salt, however, does not share properties with silver.

Cárdenas also noted that "time is very necessary . . . for the cluster of silver, ore, and mercury to be able to brew, embrace, and mix itself. Heat is also necessary for the same purpose."[83] Salt does not "embrace" anything in this process; rather, it "serves to provide heat, and activate the quicksilver, and thus helps to brew, ferment, and loosen all that ore so the quicksilver can penetrate it better, and embrace the silver."[84]

Cárdenas used a traditional framework of relationships between natural entities to explain the amalgamation process. He then offered three ways of improving it. First, he suggested replacing salt with "verdigris, or quicklime, or mercury chloride, or another hot material, yet not so hot as to consume and destroy the quicksilver." These materials "could make an effect as good as the salt, and better, if it is a hotter material." He also recommended dissolving the salt "in vinegar, or in lime juice, or in any other liquid by nature very penetrative." Finally, Cárdenas suggested that if "the ore were burnt and completely brewed, and became so pure, and thus made into powder without sand," the mercury would retrieve more silver from the ore.[85]

The first two suggestions assumed that there might be better intermediaries than salt for activating the "embrace" of silver and mercury. The only way to find those elements was through experimentation. The final suggestion concerns the need to improve mills, something that miners and other experts were already working on. (In fact, in 1559 Pedro González and Alonso de León had devised new sieves to reduce the ore to a fine powder.)[86] Cárdenas thus began with a practical problem and ended with solutions that could only be achieved through experimentation and the improvement of instruments.

The role of salt, however, was only one of the two amalgamation problems that Cárdenas discussed. The second had to do with the loss of mercury. Many miners developed methods and instruments to reduce the loss of mercury, but it was still a very costly problem. Cárdenas explains the problem thus:

> Let us assume that a miner has a pile of ore of about one hundred quintals, and that he adds to this ore thirty pounds of mercury for obtaining its silver, all of which is mixed, incorporated, and combined until extracting the silver. Afterward, when [the miner] cleans the ore, if he finds the same thirty pounds that he originally mixed, I do not mean to say that if he mixed thirty pounds of mercury, he finds the same [pounds] of mercury, rather that of the thirty pounds, if five [pounds] are of silver, the rest would be of mercury . . . this is our doubt: What could have happened to the missing mercury? In what was it consumed? . . . An even more difficult question remains: Why is it necessary that precisely and exactly the amount of mercury missing corresponds to the amount of silver extracted?[87]

Cárdenas again took a problem from the practical realm of mining and tried to find a theoretical answer to the problem.

Some people thought that the mercury was transformed into silver. "This is fantasy," replied Cárdenas, "because if this were the case, we could say that the ore did not have silver, but that the silver extracted was the same mercury transformed into silver," which is not the case. He also dismissed two other explanations, namely, that the mercury is washed away and that it sinks into the soil because of its weight. If mercury were washed away, Cárdenas explained, "black females and Indians" would find it when they looked for remnants of silver in the river—and they never found mercury there. It could not be sinking into the soil because, even when miners mixed ore and mercury in enclosures with copper planks at the bottom, the amount of mercury missing still corresponded to the amount of silver extracted.[88]

Cárdenas first attacked the problem of the missing mercury, by explaining its two important characteristics. First, mercury is "composed of very subtle and penetrative parts, so extremely thin that, at some point, fire or any other extreme [source of] heat makes them fly, and transforms them into smoke."[89] "We can experiment with" this characteristic of mercury, remarked Cárdenas, by observing that mercury volatilizes and evaporates when it is heated. Second, mercury "has the most angry adversity against fire and heat."

> This rivalry and adversity is clearly visible, because mercury . . . is very cold and very wet, and [in contrast] fire is hot and completely dry; mercury is the heaviest thing in nature . . . [while] fire is light and agile compared to all other elements. The being and nature of mercury is water, and fire is extremely contrary to it [water]. Thus all the adversity of enemies is found between mercury and fire, and, consequently, between mercury and heat.[90]

When these two characteristics are taken into consideration, "the answer to the problem" of the missing mercury "is clear": the missing mercury has been transformed into "smoke." This was easy to "see by experience" in the ovens used for the amalgamation process: when the heat was too high, the mercury, transformed into smoke, adhered to the walls of the oven.[91]

Cárdenas mixed practical observations with classical theoretical principles. Scholars like him became intermediaries between instrument-making practices and technology development, on the one hand, and the formal culture of universities and professional natural historians, on the other. In this role of intermediary, he took some elements from the culture of inventions and integrated them into his method. Testing, which in practical terms helped develop instruments and technologies, became for Cárdenas an element of his argument.[92] His relationship to artisans and their experimental approach to nature is similar to the relationship discussed previously between cosmographers and pilots.

The physician Cárdenas was not the only professional interested in these practical problems. José de Acosta, a Jesuit and natural historian, also discussed the amalgamation process in his work *Historia natural y moral de las Indias* (The Natural and Moral History of the Indies, 1590).[93] Acosta was more interested in describing the "marvelous properties" of mercury than in solving problems. After mentioning that the "ancients did not attain the secret" of using mercury for exploiting silver, he noted the ability of mercury to penetrate and consume most materials, such as iron, copper, animal skin, and clay. Mercury, Acosta remarked, also helps in making alloys of gold and copper or brass and silver. Yet, "of all the marvels of this strange liquor," the characteristic that Acosta found most interesting is that mercury,

> being the heaviest thing of the world, immediately turns into the lightest, that is, into smoke, and thus ascends, and later the same smoke, which is such a light thing, immediately turns into a thing as heavy as the same liquor of mercury into which it has turned again. [This is so] because once the smoke of that metal [mercury] touches a solid body above, or reaches a cold area, it immediately solidifies and falls down in the form of mercury. . . . This immediate transmutation of something so heavy into something so light, and vice versa, can be considered something strange in nature. In all these [characteristics] and other strange properties of this metal, the Author of its nature is worthy of glorification because all created nature immediately obeys His occult laws.[94]

Like Cárdenas, Acosta translated empirical information into theory. In Acosta's case, mercury was a marvel of nature that responded to God's laws and thus glorified the creator of the world.

Cárdenas and Acosta exemplify late-sixteenth-century theoretical approaches to nature. They both based their findings on personal experience and showed an interest in practical matters. They also drew upon empirical evidence in order to support their positions. Acosta used empirical information and arguments in addition to his classical ideas; Cárdenas used empirical information and arguments to understand events that the classical framework of elements and properties did not fully explain. Cárdenas felt that he could find causes through observation and inference; Acosta was convinced that laws of nature are manifested in nature's operations. For Acosta, therefore, the study of nature led to the understanding of the order of nature and, in turn, to the glorification of God. For Cárdenas, the study of nature led to knowledge of causes and thus to an understanding of the relationships between entities in the world.

These two cases suggest that there was, at least by the late sixteenth century, an incipient link between people working in mines, for instance, and people working in such areas as medicine and natural history. The connection between experts and professionals was also evident in the relationship between pilots and cosmographers at the Casa and between reports from explorers and official chroniclers for the Council of Indies. There were structural similarities in the practices developed in these places, such as collaboration among experts and a shared reliance on empirical reports and testing. These practices, resulting from the relationship between common people and royal officials, flourished outside academic settings. Such practices emerged in royal institutions and received the support of the crown.

By the early 1570s the empirical practices described in these chapters had already exerted a widespread influence on the way Spaniards, creoles, and the crown controlled the New World. By this time, the Council of Indies required people offering new inventions to present themselves for examination.[95] During the same years, the amalgamation process reached Peru. The physician Francisco Hernández (ca. 1515–1587), for instance, in charge of the first natural-history expedition to the New World (1571–1577), was already gathering information and samples in New Spain and collecting information on medicinal plants.[96] But the empirical culture described in this chapter was not restricted to improving mining techniques and making instruments: it pervaded most activities regarding the New World, from navigation to mining, from agriculture to natural history. Indigo seeds, along with a manual of instructions for their cultivation, reached Seville.[97] Juan Alonso de Grada (from Canarias) presented an instrument to measure longitude before the Casa de la Contratación.[98] Juan de Puy presented an instrument to purify saltwater and make it drinkable (Appendix 2, Table 8). As this empirical activity increased, it became necessary for the state to develop mechanisms for collecting and organizing information about the New World. During those years, not surprisingly, the Council of Indies began requesting regular reports about the New World.[99]

FOUR

Circuits of Information

REPORTS FROM THE NEW WORLD

Around 1533 Charles V made the following request to his officials in New Spain:

> Because we want to have complete information about the things of that land and its qualities, I order you, after having received this one [royal decree], to make a long and particular report on the greatness of that land covering its width as well as its length, and about its limits. You should write very specifically its proper names and how its boundaries are delimited and marked. Likewise, [make a report] about the qualities of the land and its wonders, particularly those of each town, and what types of native people are there, writing in particular about their rites and customs.[1]

The Spanish encounter with the New World not only fostered the development of empirical and collaborative practices to exploit, study, transform, and explore the New World but also shaped the methods used by the central state to control distant resources and lands. In particular, the crown demanded reports about the New World and came up with new mechanisms for gathering and organizing information based on those reports.

Throughout the sixteenth century the royal bureaucracy periodically requested information about the natural world of the Indies. These requests began as general inquiries about nature but over the years became very specific tools of inquiry that included precise specifications for how questions were to be answered. This knowledge emerged from reports sent from the New World and organized at the Casa de la Contratación and Council of Indies in Spain, and it enabled the crown to have maximum control over and maximum benefit from its American possessions.[2] Eventually the crown developed its own questionnaires for gathering information about the New World. These questionnaires became increasingly systematic, and by the 1570s the crown sent a printed questionnaire to the New World. From the late 1570s to the early 1580s the crown used these answers to gain an exhaustive body of knowledge about the Indies. The resulting reports are known as the *Relaciones geográficas de Indias* or, better, the *Relaciones de Indias.*[3]

During a period of about seventy years, the diffuse elements of an empirical culture came together to create unified mechanisms for gathering information and acquiring knowledge about the New World. The newly acquired knowledge depended not on traditional classical accounts of nature but on the kinds of observation and information developed in response to a bureaucratic and commercial problem: the long-distance management of natural resources.

The administration and domination of America forced the Spanish state to develop efficient ways to organize its imperial functions. In Weberian terms, the organization of the state with regard to specialized personnel as well as information collection was connected, in Spain's case, to its empire.[4] Perhaps, as the historian José Antonio Maravall argues, the empire contributed to the decline of the Spanish state. Still, in the beginning the empire fostered the creation of the central state, and together with it came the context of social values that advanced empirical practices.[5]

AN INFORMAL RESEARCH PROJECT

The *Relaciones geográficas de Indias* were the culmination of a long and unsystematic state-supported process of collecting information. From Ferdinand and Isabella to Philip II, the Spanish kings requested increasingly more information about their American possessions.[6] A good example of this type of information-gathering is the encomienda—the labor grant of indigenous American people to a Spanish colonist. To establish encomiendas, the crown requested descriptions and *pinturas* (drawings) of the land, descriptions of the population, census data, and the classification of taxable goods. Every time an encomienda changed hands, the crown requested new information. The Spanish crown used this information to introduce reforms, to make changes in tribute policy, and to design other new policies.[7]

The information-gathering associated with encomiendas and many other political ventures was informal, but it was connected with the already institutionalized project of geography and hydrography information-gathering at the Casa de la Contratación. Empirical information-gathering practices had their informal beginnings before 1520 and emerged as a result of numerous decisions and pursuits developed in the first years of activity in the New World. In those early years, no one had specific knowledge about the New World or the natural resources of the Indies. Generally, the crown assumed that the explorers would find natural products similar to those of the Old World. Thus, in 1501 the crown granted a license to Diego de Lepe, "resident of Palos," to explore the Caribbean. One of the provisions of his *capitulación* (contract) concerned trade. The crown authorized him to trade gold, silver, copper, tin, quicksilver, and other metals of "any quality," jewelry, precious stones, and samples of plants and

> animals of any quality whatsoever, and fish and birds, and spices and drugs and any other things of any name and quality even if they are of a greater value than those already listed, provided that you do not bring slaves.[8]

A contract granted to Cristóbal Guerra in 1503 contained comparable references to natural resources, but this time the list of exploitable resources included "monsters" and "serpents."[9] A 1508 contract with Diego de Nicuesa and Alonso de Ojeda included references to "anything of any kind and quality and name."[10] As these examples suggest, knowledge concerning the natural products of the Indies was still quite general and vague in the early years of the sixteenth century.

By 1508 the crown was already trying to gather more specific information about the New World. This was the result of Ferdinand II's policy toward the New World after the brief reign of Philip I. One of Ferdinand's priorities was to collect information about navigational routes. In 1508 the crown ordered Amerigo Vespucci to make a sea-chart with reports brought by pilots.[11] In 1512 (the year Ferdinand conquered Navarra to consolidate the crown's power in the Iberian kingdoms) the crown began to request particular information about the New World from explorers, as the humanist Peter Martyr had been doing at the court. Thus, the contract granted to Juan Ponce de León for the exploration and settlement of a colony on Bimini Island (in the Bahamas) included the provision of sending a report about "what is on" the island to the king and the officials of Hispaniola.[12] That year, the crown ordered the chief pilot, Juan de Solís, and the pilot Juan Vespucci—"for they are very expert" in navigation—to assemble

> as many pilots as possible and the most knowledgeable in navigation, astrolabes, latitudes, and compasses. I order those [pilots] to get together with you and thus assemble to discuss extensively the method that should be used in making a royal portolan for the navigation of all the parts of all the Indies.[13]

This royal portolan or *padrón real* was a general sea-chart. The request to Juan Ponce de León to send a report on the products and peculiarities of Bimini Island was therefore part of the emerging state mechanism for mapping and collecting information about the new lands. As mentioned above, Spaniards employed these collecting mechanisms within Spain as well. For example, in 1517 Hernando Columbus sent messengers to collect information about Spanish regions and towns for his cosmography and description of Spain. Similarly, in 1518 Diego Velázquez obtained a license from the royal officials of Hispaniola to send a fleet to Cozumel and Yucatán to "know their secrets."[14]

People on the ground were also sending their own reports. That same year, Alonso de Zauzo wrote a letter to Charles I, the recently crowned king of Spain (1516) and soon to be elected Holy Roman emperor (1519), about the natural products of Hispaniola. Among them, he mentioned brazilwood (a wood yielding a red dye), *guayacán* (guaiacum, a medicine used to treat syphilis), and fragrant resins (similar to incense).[15] As part of this emerging practice of reporting, the crown ordered Juan de Cárdenas to send a "sketch and drawing" (*traza y pintura*) of Barbados, Trinidad, and Isla Verde (east of Venezuela).[16] The crown continued to collect empirical information about the new lands and eventually established regular mechanisms for collecting information in the mid-1520s, when the Spaniards moved into the Mexica (Aztec) empire. It was then that the king formalized the administration of the New World, when he established the Council of Indies in 1524.[17]

This stage of information-gathering coincided with two other important events: the first attempt to map the nature of the New World by a natural historian and the inspection of the Casa de la Contratación in 1526 by royal officials—the year Cabot left for his expedition to the

Moluccas. The inspectors found that pilots were still using many different sea-charts for navigating to the Indies, rather than the master portolan ordered in 1508 and again in 1512. After the visit, the king ordered the best and more experienced pilots to meet with the royal pilot and establish a royal chart.[18] A few months later, on October 6, 1526, Hernando Columbus was also assigned the task of making a master portolan, a mappa mundi, and a globe depicting the new lands.[19]

The crown ordered Columbus to work with navigation experts, especially Diego Ribeiro and Alonso de Chaves. During this period, Ribeiro made some maps of the New World. The following year, the crown also ordered pilots and shipmasters to send reports from their voyages with indications of the routes, lands, and islands found as well as distances, latitudes, and descriptions of the coasts. This royal decree marks the beginnings of the crown's active interest in gathering empirical information about the New World's geography and the mapping of the New World.[20]

In 1526 Gonzalo Fernández de Oviedo wrote and published the first book on the natural history of the Indies, the *Sumario de la natural historia de las Indias,* at the request of Charles V. He wrote it from memory, for he had left his notes and writings in Santo Domingo. It was not a coincidence that at the time of Oviedo's presence at the court in 1525 the crown ordered its officials in Hispaniola to send numerous curiosities to the humanist and royal secretary Peter Martyr d'Anghiera:

> Parrots, turkeys from Tierra Firme, and other strange birds, iguanas, spices of chili pepper, cinnamon, cañafistula, or other fruits, or roots of trees, or blue or amber stones or anything.[21]

Oviedo's testimony portrayed the nature of the Indies not as a collection of commodities but as a collection of wonders. And so it was that wonders joined the trans-Atlantic circuit of commodities. It was during these years that Ribeiro obtained his license for his bilge pump and Villasante presented his report on balsam before the Council of Indies.

The crown was not as interested in collecting marvels as in gathering information about the new lands and their commodities. In 1530, for instance, the king ordered the president of the audience (a tribunal and its territory) of New Spain, Don Sebastán Ramírez de Fuenleal, to send reports about the land and its inhabitants. This process would eventually produce great geographical knowledge about the New World. In 1532 the crown added more formal mechanisms for collecting information. First, the crown requested reports from merchants, officials, and explorers in the field, acknowledging this method as the best way to create an official description of America and its natural resources. Second, the crown incorporated those reports into the structure of the imperial state for collecting information. Thus, while the crown continued to request reports from particular individuals, it also began to send orders to royal officials requesting information on the new lands. In that year, the crown appointed an official—not surprisingly, Gonzalo Fernández de Oviedo—to collect and organize information regarding the natural history of the New World.[22]

REPORTS FROM THE NEW WORLD: A HISTORY

In 1532 Oviedo was appointed chronicler of the Indies ("cronista de Indias") and charged with writing their social and natural history. The king specified that "whatever he writes could not be published without first giving a copy to his Majesty for its examination";[23] he gave Oviedo an annual salary of 30,000 maravedís to write "the things of the Indies."[24] This was 20,000 maravedís less than Amerigo Vespucci's salary as chief pilot, but it was still good remuneration for the first self-fashioned royal natural historian of the Indies.

The crown thus dispatched royal decrees to its officials in the Indies requesting information about natural history. The royal decree sent to the governor of Fernandina Island (Cuba) asked for information "about the island, its dwellers, and its conditions."[25] It explained that Oviedo was writing a general history of the Indies as well as a natural history of the lands and islands, its animals, and "its strangeness." The decree also mentioned that every year Oviedo would have to provide a copy of his own writings to be added to the history of Spain. The governor would therefore have to send information, as promptly as possible, every time it was requested. In all cases, these reports would have to be signed by the people who provided them—a significant requirement, because it cast individual testimony into the circuit of knowledge-gathering practice.[26]

With Oviedo's appointment, institutionalized mechanisms for obtaining natural history information about the New World were set into motion. The significant feature of the royal decree of 1532 was its concern for information about the natural world of the Indies—about strange things, animals, and characteristics of the land. By contrast, a previous royal decree in 1530 (to the audience of Hispaniola) focused on urban and social matters—it requested information about types of towns, married and unmarried residents, ports and their officials, royal officials in general, fortresses and houses, churches, their benefices and who held them, and the quality of the land.[27] A 1527 decree emphasized geographical information for the development of maps. But by 1532 the crown's needs had changed—the Santo Domingo balsam was already selling in Spain, the king was receiving curiosities such as boxes of medicines and chiles from Santo Domingo,[28] and Oviedo had already published his *Sumario*. The ecological transformation of the New World—to make it habitable for Europeans, but, more important, to make it economically feasible for them to settle or conduct business there—was already well underway. The natural world of the New World being discovered had already been changed in some ways from its pre-encounter state.

It was also in 1532 that the crown commissioned the cosmographer Alonso de Santa Cruz (ca. 1500–1572) to produce new navigational charts. He needed information about the "grades of the lands and islands." The crown sent a royal decree to the officials of the Casa de la Contratación that required shipmasters and pilots to give Santa Cruz all the information he needed.[29] A few years later, Hernando Columbus received a royal decree admonishing him to finish the navigational chart that he had been ordered to make in 1526.[30]

Between the mid-1520s and early 1530s the crown became aware that the mapping of the natural products of the Indies provided not only economic opportunities but also the basis for a policy of good imperial government. Understanding and studying America's resources became part of imperial policy aimed at dominating and controlling distant lands. This practice of empirical information-gathering emerged within the context of mercantilism, commercial exploration, and imperial activities.

In the early 1530s a new chapter in the information- and knowledge-gathering practices relating to the natural world of the Indies began. Santa Cruz's royal chart, Oviedo's natural history and his appointment to the Council of Indies, and the royal decrees requesting information about natural history and natural resources all appeared in this decade. The first impulse to map lands and catalog natural resources came from people in the field who needed this information. The crown followed suit and implemented formal information-gathering programs.

Reports from pilots and explorers were increasingly precise in their descriptions. In a 1534 report about the exploration of the Pacific Ocean, one pilot (presumably Martín de Acosta) gave almost daily accounts of his navigation, detailing altitude, distances, and directions. Hernán Cortés, who financed this venture, may have requested that the pilot (Martín de Acosta) as well as the captain (Hernando de Grijalva) write this report. Cortés himself, in an undated letter to the king, suggested that explorers should report on the lands, on indigenous populations and their rites and customs, on natural products, and on the characteristics and latitude of ports.[31] The report describes the coast and islands visited by the expedition as well as marine creatures like one that "came very close to the ship." The pilot explains that he and his crew "were unable to say what type of fish it was. Some said it was a sea man, others said it was a sea wolf."[32] Days later, in a different area, they spotted a similar "fish."

The anonymous pilot made a drawing of the creature and included it in his report. Because he was not sure of what he had seen, he drew two versions of his "fish": one with scales and the other without.[33] It is possible that what they saw were sea-lions. The pilot appended his reports on distances and grades with empirical information about the lands and products that he had found.

The process of gathering information, together with all invasions and explorations, slowed down during the 1540s, when the crown was listening to the Dominican Bartolomé de las Casas (1474–1566) and the theologian Juan Ginés de Sepúlveda (1490–1573), who would argue their positions on the difficult topic of the Amerindians in the great debate of 1550 at Valladolid.[34] Religious tensions were on the rise in Europe: John Calvin was back in Geneva implementing Protestantism; Pope Paul III established the Company of Jesus (Jesuits); and the Catholic clergy in Spain began a process of reform.[35] During the 1540s, Charles V was directing all his attention toward the Protestants in the Holy Roman Empire, after some years of fighting the Muslims (he tried to take over Argel in 1541 but lost).

Nevertheless, there was still interest in the New World. Oviedo printed the first part of his *Historia natural y general de las Indias* in 1535, with a second edition in 1547. The second part was printed in 1557 (the complete edition appeared only in 1851–1855). There were a few decrees in the 1540s and 1550s requesting geographical information as well as data on natural products.[36] In 1552 Francisco López de Gómara published *La istoria de las Indias,* compiling information from Hernán Cortés and others involved in the Mexican adventure. He provided some information about animals and geography, with an illustration of a bison and a map of the New World (see Figures 4 and 5).[37] In the mid-1550s and 1560s a renewed interest in the natural world of the Indies began to emerge among royal officials.

4. *Bison in Francisco López de Gómara,* La istoria de las Indias, y conquista de Mexico *(Zaragoza, 1552). Reproduced courtesy of the John Carter Brown Library at Brown University.*

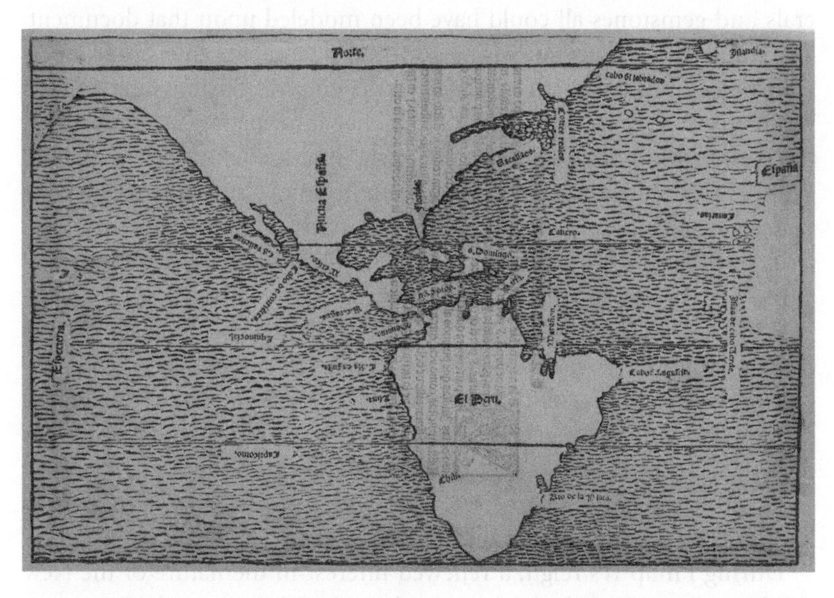

5. *Map of the New World in Francisco López de Gómara,* La istoria de las Indias, y conquista de Mexico *(Zaragoza, 1552). Reproduced courtesy of the John Carter Brown Library at Brown University.*

The first foray of this new phase was the *Memorial* of Alonso de Santa Cruz, written for the king in the mid-1550s, presumably just a year after the ascension of Philip (1556).[38] It contained questions for New World explorers regarding the latitude and longitude of places and ports; the geographical characteristics and healthiness of the land; descriptions of rivers, mountains, lakes, and fountains; and information about mines, minerals, stones, pearls, animals, "monsters," trees, fruits, spices, drugs, and herbs. Finally, there were questions concerning the indigenous people—their kingdoms and provinces, borders, towns and cities, costumes, rites, types of knowledge, books, arms, trade (in general), and items that they traded.

Santa Cruz not only drew upon his knowledge and experience as a cosmographer in creating this questionnaire but also exploited the resources of his own personal archive, the "papeles de Santa Cruz" (Santa Cruz's papers). Among his papers was the anonymous pilot's document (cited earlier in this chapter) about the expedition to the Pacific Ocean. Santa Cruz's questions about geography, the location of places as determined by latitude, waters and lands, strange animals and "monsters," and minerals and gemstones all could have been modeled upon that document. The similarities are not surprising, for these documents emerged from established empirical practices related to the empire. Cosmographers used mechanisms such as these reports to gather information not only about the New World but also about previously little-known places in the Old World. For instance, a report about Japan in Santa Cruz's possession contained information organized around questions similar to those in Santa Cruz's *Memorial.*[39] Not surprisingly, Santa Cruz applied his skills to prepare a map of Spain. Santa Cruz noted:

> I have completed a map of Spain, more or less the size of a large tablecloth [*repostero*], showing all the cities, towns, and villages, the rivers and the mountains, together with the frontiers of the kingdoms and many other details.[40]

Nothing else is known about this map.[41]

During Philip II's reign, a renewed interest in the nature of the New World and, incidentally, in the mapping of Spain appeared. The crown launched three major geographical projects in Spain. Phillip commissioned Anton van Wyngaerde (ca. 1512–1571) to make topographical drawings of Spanish towns (beginning in 1563); the crown asked the geographer Pedro de Esquivel (d. 1575) to produce a geodesic description of

Spain (mid-1560s); and, finally, Juan de Ovando sent a questionnaire to Castilian communities to gather specific information (ca. 1574).[42]

Pedro Arias de Benavides published his book on New World medicines, *Secrets of Chirurgia*, in 1566.[43] He described his experiences as a surgeon in Guatemala and Mexico, the herbs and medicines found in the New World, and their uses by the indigenous people. Tomás López Medel (1565–1571) began writing his own treatise on the natural world of the Indies, perhaps at the request of Juan de Ovando. His work was the first attempt to present the nature of the New World within the classical categorization of the elements and their properties.[44] These activities provided the framework for the final step in the institutionalization of information-gathering, as promulgated in a series of statutes for the Council of Indies in 1571 and 1573.[45]

THE COUNCIL OF INDIES: COLLECTING INFORMATION

The man behind the institutionalization of information-gathering and knowledge acquisition at the Council of Indies, Juan de Ovando (1515–1575), came from a family of peasants, attended the college of San Bartolomé at Salamanca, and became a lawyer. After serving for some years in the archdiocese of Seville as inquisitor, he was appointed to reform the University of Alcalá (1564–1566). By that time, Ovando was an ally of the powerful cardinal Diego de Espinosa (president of the Council of the Inquisition and later president of the Council of Castile). After Alcalá, Ovando was appointed to the Supreme Council of the Holy Inquisition. Later he was appointed to inspect the Council of Indies (1569). From 1571 to 1573 he was the president of the Council of Indies; afterward he became president of the Council of Finance (1573–1575).[46]

During his inspection of the council in 1569, Ovando reformed its legislative and administrative activities and established more systematic information-gathering practices. Years later, the chief chronicler of the Indies, Antonio de Herrera (appointed 1596, d. 1624), explained that Ovando ordered the collection of "the most accurate reports found in Spain and the Indies about the events of the discovery of the Indies," to address the lack of information and uncertainty at the council regarding that matter.[47] Ovando's visit to the council resulted in the formalization of a program for collecting information about the nature and geography of the Indies.

Before the 1570s the Council of Indies had been indirectly involved in some of the navigational activities of the Casa. A member of the council,

Juan Suárez de Carvajal, drew up the 1536 statutes for the Casa, which organized the examination of pilots and the development of the portolan. The council also appointed the pilot and cosmographers recommended by the officials of the Casa; however, after Juan de Ovando's inspection, the council itself became more engaged in a wide range of scientific practices, including natural history, geography, hydrography, cosmography, and scientific expeditions.

The recorded meeting minutes declaring the motives behind Ovando's inspection are lost; but at the time of his appointment, government officials were widely dissatisfied with the viceroys in Peru and Mexico. Ovando's task was to investigate the council's activities (e.g., appointing royal officials in the New World). He identified two major problems. The first was the lack of a system for reporting information about the New World. As noted above, reports about the New World came from private initiatives requesting licenses or grants from the crown or from royal economic and political initiatives. There had never been a systematic attempt to collect the same kinds of information in a consistent manner from everywhere in the New World. Second, Ovando found a lack of uniform legal structure in the American kingdoms.[48]

Ovando's report, and his presidency, precipitated the statutes of 1571, the *Ordenanzas reales del Consejo de Indias.* Among other changes, the statutes created the new office of the chief cosmographer-chronicler, which was filled by Juan López de Velasco (1530–1598). López de Velasco's duties consisted of writing the history of the Indies, censoring books about the New World, and collecting information about geographical and natural resources.[49]

More generally, the statutes of 1571 established the following provision:

> Since nothing can be understood or appropriately dealt with if its subject is not first known by the people that would have to know and decide about it, We order and command that those [officials] of Our Council of Indies endeavor with particular study and care always to have a complete and certain description and inquiry about all the things concerning the condition of the Indies, from land to sea, natural and moral [things], eternal and temporal, ecclesiastical and secular, past and present, and those [things] that in time could fall under government or legal jurisdiction.[50]

One example arising from this directive was the 1572 royal decree to the president and officials of the Audiencia of Quito, asking them to obtain information "from any people, secular or religious," about discoveries and wars; about the native peoples' religions, customs, and rites; about the land; and about nature and the "qualities of the things" within the province. It also requested that documents from government archives regarding these topics be sent to the newly appointed officer "in charge of compiling and writing the history of those [lands]."[51] This decree echoed Oviedo's 1532 decree, which had been the result of his own proposal and the council's support; López de Velasco's decree was the result of a council directive. The collecting of information about the New World was now part of the bureaucratic structure, and the statutes of 1571 were just the beginning of this bureaucratic expansion.

Similarly, the 1573 statutes for new discoveries and settlements established that explorers should make daily reports and descriptions of "what they see and find, and of what happens during the discovery." All this was to be written in a book that "should be read in public to better determine the truth."[52] Recall the anonymous pilot's report of 1534. That report, and others like it, provided precedents for defining such institutionalized norms regarding information-gathering and consensual inquiry.

In 1573 the crown also issued statutes for the formation of a book describing the Indies. These statutes outlined a number of stipulations. First, they dictated the status of the people who should engage in writing the book; second, the "things about which it is necessary to make description, inquiry, and report"; and, third, the method and manner of the descriptions.[53]

The officials in charge of making the descriptions were almost all engaged directly in the American enterprise. They included high-ranking members of society such as officials of the Council of Indies and the Casa, archbishops, bishops, clerics, viceroys, presidents and officials of audiences, governors, mayors, council members both Spanish and indigenous, caciques, and treasury officials. They also included those with particular knowledge about natural resources of the New World, such as captains and fleet admirals, pilots, shipmasters, and captains of provinces. Even common people were ordered to write reports:

> Any of our subjects and vassals who would reside and now reside [there] or who travel to any part of the Indies and who know and understood its things, we entrust and command [them] to make certain and true reports.[54]

The Book of Descriptions, as it was called, covered all aspects of the New World under the following categories and subcategories: cosmography (climates, longitudes, latitude, eclipses, celestial marks); hydrography (coasts, longitude and latitude, positions and characteristics of rivers, ports, annual weather conditions); natural history "constantly of each region" (domestic and wild animals, uses of animals, hunting methods, breeding practices, types of fish, uses of fish, methods of fishing, wild and domestic birds, hunting methods, useful birds, breeding methods, trees, plants, crops, types of wood, fruits, domestic and wild herbs, useful herbs, metals, types of land); and moral history (discoveries, conquests, peoples, their kingdoms, tributes, services, religion, customs, rites, food, communal and private belongings, types of contracts, crimes, punishments, kings, types of government, arts and trades, wars, arms, writing systems, calendar systems, Spanish communities, cities, houses, streets, lands). Finally, the Book of Descriptions also included provisions about administrative and legal matters regarding the New World. Royal decrees were dispatched to New Spain, New Galicia, Hispaniola, Guatemala, Panama, Quito, the New Kingdom of Granada, Chile, and Charcas with the order to implement the book's statutes.[55]

As different and more complex issues were raised through the experience of long-distance domination, the crown was compelled to systematize the information coming from America to keep it under its control. The next systematic attempt to do so occurred with the reports of 1577.

THE *RELACIONES DE INDIAS*

Juan de Ovando himself had already begun the process of systematically collecting information about America. In 1569 Ovando circulated in the Indies a questionnaire of 37 chapters. In 1571, upon his advice, the Council of Indies circulated a questionnaire of 200 chapters to those coming from America—as the Casa pilots and cosmographers had been doing for years.[56] In 1573 came yet another questionnaire of 135 chapters.[57] The 200-chapter questionnaire was the basis for the 50-chapter questionnaire administered in 1577, which was reprinted with minor changes in 1584.[58]

The political purpose of the reports was laid down in the title of the questionnaire, "Instruction and questionnaire for the reports that must be done for the description of the Indies, commanded by His Majesty, for their good government and ennobling."[59] The questionnaire was sent via viceroys or audiences (who had to answer the questionnaire as well) to governors, magistrates (*corregidores*), and mayors. These officials were instructed to draw up lists of all Spanish and Indian towns under their jurisdiction and to send the "Instruction and questionnaire" to these towns. In each town, the questionnaire was to be sent to the municipal council members. If there was no council, it was to be sent to a priest (or to a friar if the town had no priest). The "Instruction and questionnaire" had to be sent back with the reports to the official who had sent it, for further distribution in manuscript form, among towns that had not received it.

The government officials who received the questionnaire could either answer it themselves or find "people knowledgeable of the things of the land" (*personas ynteligentes de las cossas de la tierra*) to answer it.[60] In general, more than one person contributed to the completion of the questionnaire. A notary testified to the identity of the informants and the authenticity of the reports. At every stage, the bureaucrats who drew up the questionnaire and those who responded took careful steps to authenticate and certify the truth of the answers.

The questions had to be answered in order, leaving blank those for which there was no known answer. The answers had to be short and clear, "stating as certain what was so, and as dubious what was not certain; thus, the reports would be certain in conformity with the questions asked." The definition of certainty was based on the personal experience of the "people knowledgeable of the things of the land" and on consensus. This process was similar to the one already established at the Casa and in the statutes of new discoveries (i.e., the reading of the information "in public") "to better inquire into the truth."[61] Finally, a complete report would contain both a written text and a drawing of the town and its setting.

The first forty-eight chapters were divided into four groups: Chapters 1 to 10 concerned towns settled by the Spanish; Chapters 11 to 15 were devoted to Indian towns; Chapters 16 to 37 were general chapters for both Spanish and Indian towns; Chapters 38 to 48 concerned ports and coastal towns and towns abandoned by the Spanish. Finally, Chapter 49 inquired about other "notable things in nature, and any effects of earth, air, and heavens presented in any place whatsoever that are worth mention." Chapter 50 asked that informants sign the report and forward it, together with the "Instruction and questionnaire," to the authority that had dispatched it to them.

The chapters concerning Indian and Spanish towns requested similar information from each: names of towns and their meaning; history (for Spanish towns, their foundation history; for Indian towns, their history before Spanish domination); weather (temperature, rains, wind patterns); descriptions of territory (latitude, distance to the closest viceroyal court or Audiencia, distances to nearby towns, condition of roads). Chapter 5 asked for the number of Indians in the area, whether or not the Indians had been dying, and, if so, how many had died and the cause of death.[62] Chapter 14 asked about Indians' rites, costumes, and tributes.

The general chapters requested information about the towns' natural resources and surroundings. They inquired about the location of the towns; their names; whether the location was healthy or deleterious; distances to mountains and their names; distances to rivers, their characteristics, and anything worthy of mention about them; distances to lakes, lagoons, and springs; volcanoes, caves, and "things in nature notable and worthy of being known"; types of wild trees, fruits, and wood; the variety of native and imported Spanish trees and their fruits and the kinds of Spanish trees that were unable to be cultivated there; types of grains and vegetables eaten by the natives and those brought from Spain; herbs and medicinal plants; animals, both wild and domesticated, whether native or brought from Spain; mines, minerals, and gems; salt-mines or other sources of salt; fortresses and ports; economic activities and tributes; religious jurisdictions; churches and benefices; monasteries and their history; and, finally, hospitals and schools and their history.

In general, the informants were long-term residents (Indians, mestizos, or Spaniards). For instance, the informants in the report from the *corregimiento* (jurisdiction of a magistrate) of Abancay (Apurímac, Peru) were Pedro de Plascencia, seventy years old "more or less"; Juan de Luque, seventy-three years old "more or less"; and Francisco Gallegos, age thirty-six. Plascencia and Luque, both Spaniards, had lived in the area for more than forty and twenty years, respectively. Gallegos was a mestizo who said that he had grown up in the area. The *corregidor* (Spanish magistrate) Niculoso de Fornee convened them, and they declared upon oath "to tell the truth."[63]

To complete the report from the *repartimiento* (encomienda jurisdiction) of San Francisco de Atunrucana (Peru), the *corregidor* Luis de Monzón summoned the priest Pedro González and Juan de Arbe, two "Spaniards who have lived for a long time in this *repartimiento.*" To obtain information regarding Indian towns, Monzón assembled the translators Juan Alonso de Badajoz (mestizo), the above-mentioned de Arbe (a Spaniard who "understands the language" of the Indians), and "Don Juan Guancarilla, Don Cristóbal Auchuqui, Don Francisco Hernández, Don Francisco Curiaymara, and other *curacas* [chiefs] and Indian notables."[64]

In some cases the report seems to have been produced by one person. The report from Atlitlalaquia (Mexico) was perhaps made primarily by magistrate Valentín de Jaso. He wrote the answers, but he did receive help from others.[65] Some reports contained little input from native informers. Although native informants apparently participated in preparing the reports from Mexicaltzingo, Culhuacán, and Iztapalapán (Mexico), the content of these reports is so poor that, according to the scholar René Acuña, "it is possible that the role of the native informers was a little less than secondary."[66]

The rate and speed at which recipients responded to the questionnaires varied. The responses arrived from Venezuela between 1578 and 1579; from New Spain, between 1579 and 1581 and 1584 and 1585; from Ecuador, in 1592; from Peru, in 1583 and later between 1585 and 1586.[67] The period from 1577 to 1586 was one of intense change for the Spanish empire. In 1579 the Moriscos rebelled at Alpujarras; in 1580 Portugal became part of the Spanish empire; in 1583 a new Spanish assault against the Protestants in the Netherlands began. Internally, the process of Tridentine Catholicization of laymen and laywomen (the implementation of practice and doctrine established for the Catholic Church at the Council of Trent, 1545–1563) was almost completed. The Counter-Reformation was becoming a cultural reality in Spain.[68] In this context of rapid external and internal change, the *Relaciones* (the answers to the 1577 questionnaire) arrived at the Council of Indies and became part of the archive at the disposal of the cosmographer-chronicler.

Parallel to the 1577 questionnaire, the cosmographer-chronicler López de Velasco developed a set of instructions for the collection of information about lunar eclipses and to determine (with the information collected) latitudes and longitudes of sites in America. Once again, his instructions were the result of the statutes of 1571. Chapter 117 ordered the cosmographer-chronicler

> to produce and organize cosmographical tables of the Indies. He must register, according to the rules of geography, the longitude, latitude, and distances of provinces, oceans, islands and mountains, and other locations usually drawn.[69]

López de Velasco drew up the "instructions for the observation of lunar eclipses . . . and ascertained thus the latitude and longitude" of various locations in America, which were in circulation after May 25, 1577.[70]

The questionnaires and instructions of 1577 articulated two overlapping and complementary tendencies in the territorial representation of America: first, there was the mathematization of the New World; second, the "regional detail of chorography" (describing or mapping the region) for the description of America.[71] A "cosmographic revolution" was taking place at the turn of the century, when richly described marginal places (such as Brazil and New Mexico) were brought together with a center (Europe as the empire) in a geodesic representation of the globe.[72] This ethnocentricity in the representation of the earth was also a by-product of the American experience. During the sixteenth century, this cosmographical program was Christianized to serve the aims of natural theology. Great cosmographers like Sebastian Münster, Joachim Vadianus, Josiah Simler, and Jodocus Hondius transformed the map into an expression of the "design of divine providence."[73] In Spain, the cosmographical program was designed to serve the aims of the empire and the crown and culminated in a cosmographical expedition to the New World in 1583—the same year that the Academy of Mathematics of Madrid began its lectures.[74]

A COSMOGRAPHICAL EXPEDITION

In 1583 López de Velasco proposed to the Council of Indies that a cosmographical expedition to the New World be carried out by Jaime Juan, "a mathematician." Velasco defined the tasks of the expedition as follows: first, to "take the altitude or elevation of the places where he would go"; second, "to ascertain the deviation of the compass in relation to the pole in the said places"; third, "once this deviation is found, to come to know the longitude and the east-west navigation by means of the instrument of longitudes that he carries with him"; fourth, to "observe the eclipses of the moon that would happen in order to find out the longitudes and distances from province to province"; fifth, to "make inquiry into and report of the times and hours of the high and low tides of the sea in the coasts and seas where he would go"; and sixth, to organize the observation of lunar eclipses in Spain and America. Finally, López de Velasco explained that "the accuracy and precision of the instruments and the intelligence of this said Jaime Juan in the use of them cannot be judged without looking at and examining them. But if those instruments were those made by Juan de Herrera, chief master of architecture of His Majesty, they can be well taken as certain and well designed."[75]

The council found Jaime Juan, "from Valencia, a man expert in mathematics and calculations in astronomy," appropriate for the expedition. It commanded him to go first to New Spain and then to the Philippines. In New Spain he was ordered to meet the cosmographer Francisco Domínguez and in the Philippines to collect the papers of Fray Martín de

Rada. The council expected the expedition to last between six and eight years and assigned Juan an annual salary.[76] Besides the required calculations and observations, the council also requested that Juan "attempt to make maps, in particular of those lands and provinces where he would go . . . and make a separate report about the noteworthy things of said provinces."[77]

By 1584 Juan was already in New Spain making observations. He prepared a report on the eclipse of November 17, 1584. Several people helped him: Gabriel Gudiel, Francisco Domínguez (who had helped the royal physician Francisco Hernández during his expedition, 1571–1577), and the physician Agustín Farfán. When Jaime Juan died, presumably in late 1584 or early 1585, the expedition came to a halt. Years later, the council would cite the aborted expedition to convince the king to support another one.[78] From Oviedo's proposal of a natural history expedition in 1525 to Hernández's medical expedition of 1571 to 1577 to Juan's cosmographical expedition of 1583, expeditions became one more tool in the state-sponsored program to collect empirical information about the New World.

These questionnaires and expeditions implicitly established a program for gathering information about the material culture of the Atlantic world. Nature was only one aspect of this emerging material culture. A similar questionnaire or program of research for travelers was established by the Royal Society of London a century later, in 1665.[79] The Spanish questionnaires show the significance of personal experience for the systematic collection of information about the New World. There were certainly classical influences in the framing of the questions; for example, those about savage and domestic animals or wild and domestic plants were likely derived from Pliny's well-known *Natural History.*[80] Nevertheless, it was in the sixteenth century that a program of gathering empirical information and acquiring knowledge about the natural world was established for the first time outside of universities and humanist circles for the study of natural history. These classical influences were at most a framework and springboard for the development of Spain's apparatus of long-distance control based on empirical practices and empirical knowledge.

The empirical and collaborative orientation of these reports, in contrast to traditional examples, evidences a complete lack of concern for Aristotelian final and formal causes. Nor did these reports contain any hint of a teleological, Aristotelian view of nature.[81] Likewise, the practice of empirical observation showed that the crown had no interest in the kind of invisible and Platonic agencies inherent in the magical and Neoplatonic elements of Renaissance thought. Pilots, artisans, merchants, and royal officials argued that those with practical experience (rather than theoretical knowledge alone) understood the New World best and developed increasingly more and better knowledge.

Within some sixty years, the Spaniards moved from a general understanding of the natural world of the Indies—a vague collection of "things of any name and quality"—to a complex and diverse understanding of nature. In a collective process of improvisation and design, the crown institutionalized, first at the Casa de la Contratación and later at the Council of Indies, the appropriate empirical mechanisms and practices for dealing with a world that no longer fit the image of the classical world. The early Scientific Revolution resulted in the questionnaires and expeditions discussed in this chapter; yet this empirical culture emerged among scholars and collectors as well. Even individuals without royal support collected curiosities from and wrote natural histories of the New World. Their story comes next.

FIVE

Books of Nature

SCHOLARS, NATURAL HISTORY, AND THE NEW WORLD

On March 12, 1579, bad weather hindered the Spanish fleet from leaving for the New World from the port of Sanlúcar de Barrameda. The day before, the fleet had attempted to leave and was forced to return. One ship was wrecked; and forty women and ten or twelve men drowned. Another ship was damaged but able to return to port.[1] The newly appointed bishop of Tucumán, Fray Francisco de Vitoria (1578–1584; d. 1592), was aboard one of the ships and wrote to the king about the situation, commenting: "necessarily if God does not provide a little bit of better weather, [the ships] are in great danger and tossed like dice."[2] In this view of the natural world, nature was chaotic and unpredictable. When nature's inherent order was disrupted, God would put things back in order—so Bishop Vitoria believed.

The bishop and the king belonged to a learned community, and their ideas about nature came from religious and classical texts, such as Genesis and the work of Pliny and Aristotle. Yet some members of this learned community, such as Bishop Vitoria, traveled to the New World; there their ideas about nature would be tested in a new environment. This chapter discusses the approaches to nature fashioned by people of learned communities who came into contact with the New World (see Appendix 3). Not surprisingly, these approaches combined humanist and Aristotelian theories with empirical methods that emerged through the context of commercial and imperial expansion and had much in common with the approaches of merchants, artisans, and bureaucrats discussed earlier in this book. Employing these empirical methods was the very first step in the reconfiguration of scientific practices taking place in the European-Atlantic world: Copernicus's ideas would become relevant after Galileo Galilei collected empirical evidence in the 1610s. By the early 1610s there was already an empirical culture that validated personal experience as a source of gaining knowledge, affiliated with Spanish institutions and books that had reached Galileo's Italy and Bacon's England. The empirical practices discussed thus far were connected to the Spanish encounter with the New World.

As Europeans arrived in these strange lands, they could not avoid feelings of admiration, a mixture of aesthetic feeling and economic possibility. "Hispaniola is marvelous," exclaimed Christopher Columbus:

> the hills and the mountains and the fertile lowlands and the countryside and the beautiful and fertile lands [are] appropriate to sow and plant, and to raise cattle of all sorts, [and] to build villages.[3]

The humanist and royal counselor Peter Martyr,[4] the royal official Alonso de Zauzo,[5] the notary-turned-conqueror Hernán Cortés,[6] the royal official and natural historian Gonzalo Fernández de Oviedo,[7] the physician and merchant Nicolás Monardes,[8] and others echoed these feelings. It was at once a unique new land full of wonders and a familiar natural world full of commodities. The notion of nature as commodity prevailed among the community of royal officials and merchants, while the notion of nature as a source of wonder prevailed among the community of physicians and natural historians (that is, the community of authors who wrote for publication).[9]

In the American context, empirical information was a way of explaining natural events. Perhaps the single most important difference between the development of natural history in the Atlantic world and in the Old World was that the natural products of the Atlantic world lacked a reference in classical traditions. Not a single classical or religious text could provide information on an avocado or cochineal (an organic red dye). Sometimes the texts did provide a clue, but it was never specific enough. The manatee is a good example of this: observers decided it was a male siren; and although classical authors had much to say about sirens, they had never dreamt of a male one.

When texts did provide actual information, they were incorrect. The question of the inhabitability of the Tropics was much discussed both by pilots with New World experience and by philosophers, poets, and cosmographers, who relied on classical texts.[10] In the 1520s Peter Martyr could not decide between one group and the other; by the 1580s, in contrast, the Jesuit José de Acosta (1540–1600) could embrace the side of experience without hesitation. Acosta excused Aristotle's ignorance on the topic of the Tropics (he claimed that life was impossible there) and moved on to refer to his own experience.[11] Similarly, for Fray Pedro Simón (1574–ca. 1626), "the experience of seeing the New World inhabited shows that this [idea of the inhabitability of the Torrid Zone] had been a false [idea], [as has been the case] with many others [ideas] held by Aristotle."[12]

The disparities between classical knowledge and New World experience forced a slow reorganization of existing epistemological models. A significant feature of these intellectual developments was the challenge faced by Europeans who sought to integrate the new information into their own cultural systems. By the sixteenth century, these cultural systems had already entered a phase of dynamic change, particularly after the recovery of lost classical texts by the humanists. With the increase in the variety of books and the availability of original versions, the process of eroding the authority of the printed word had already begun. The information from the New World helped to accelerate this process. What was left from the ancient authorities was a set of frameworks to organize the new information.

The task of natural historians and physicians was to adjust existing cultural models to fit new realities.[13] Physicians, for instance, placed their books within the framework of elements (earth, water, fire, and air) and qualities (hot, cold, dry, and wet), but they described a Torrid Zone where people wore alpaca-wool sweaters high in the Andes. Once they incorporated empirical information, they opened up their texts to the concept of allowing personal experience to serve as an independent source of knowledge. Empirical information therefore partially displaced classical authorities and increasingly became the prevailing source of relevant information about the New World among scholarly communities.

Natural historians, too, transformed the New World into a familiar one by framing its commodities and wonders within classical theories about nature. This is particularly evident among the community of collectors in Seville and in the primary natural histories published in the sixteenth century: the books by Gonzalo Fernández de Oviedo, *Sumario* (1526) and the *Historia general y natural de las Indias* (first part published in 1535 and 1547),[14] and José de Acosta, *Historia natural y moral de las Indias* (1590). These histories marked yet another attempt by European scholars to interpret experience through classical theories prior to the great break with those theories in the seventeenth century.

THE DIVERSITY OF NATURE

Gonzalo Fernández de Oviedo (Madrid, ca. 1478–Santo Domingo, Hispaniola, 1557) was the first scholar to study the New World's nature. His travels, contacts with humanists, and knowledge of court systems had fortuitously prepared him to fashion himself as a natural historian once he came into contact with the New World. After arriving at this unexpected destiny, this self-fashioned natural historian wrote the first natural history of the New World, *De la natural historia de las Indias* (1526), known also as the *Sumario de la natural historia de las Indias.*

Oviedo admired the nature of the New World:

> What mortal mind would be able to understand such diversity of languages, customs, and habits in the people of the Indies? Such diversity of animals, both domestic and savage? The indescribable multitude of trees, [some] abundant with diverse types of fruits and others sterile; [trees] cultivated by Indians, as well as those that Nature, by its own operation, produces without the help of mortal hands? All the plants and herbs useful and beneficial to mankind? All the other innumerable [plants and herbs] unknown to him [man], with their different roses, flowers, and odorous fragrances? All the diverse birds of prey and other kinds? All the high and fertile mountains and the others so different and wild? All the fertile and arable lands ready for cultivation and with good riversides?[15]

With these rhetorical questions, Oviedo sought to transmit his feelings of admiration for the New World and at the same time to justify his book of nature. His questions posed two problems: how to describe the diversity of nature in the New World and how to explain this diversity.

Oviedo wrote his *Sumario* from memory, for he had left his notes in Hispaniola. (They would later become the basis for the *Historia general y natural de las Indias.*) Oviedo's *Sumario* discussed the navigation to the New World and described customs and ways of the indigenous people, land animals, birds, rivers, streams, seas, fish, plants, herbs, "and things that produce the land."[16]

The *Historia* expanded upon these themes in terms of detail and additional information. Between the *Sumario* and the *Historia*, Oviedo described more than 250 animals and plants.[17] The *Historia* also included administrative and military information about the Indies, for "although this [information] is not part of the natural history, it is necessary to understand the principle and foundation of everything."[18] Both texts are part of the same understanding of nature, and they are treated here as a single work. The *Sumario* was intended as the first report about nature in the New World. Previous reports (such as Peter Martyr's *Decades of the New World,* or the letters of Columbus, Vespucci, and Cortés, or Martín Fernández de Enciso's *Suma de geografía*) did not deal exclusively with the natural world of the New World, as did Oviedo's text. During the course of the sixteenth century, his text was translated into Italian, French, and English.

As mentioned above, Oviedo had prepared himself to become a natural historian through his travels and contacts with the humanist culture of the Spanish and Italian courts. Oviedo "traveled around all Italy," where he did all he "could to learn and read and understand Tuscan and to find books."[19] He also became the best (according to his own account) in the art of cutting paper figures with scissors, without drawing lines, to entertain his courtly patrons.[20]

Oviedo's ability to amuse courtly patrons, among other things, helped him to move from the court of the Duke of Villahermosa (1490) to the court of Prince Juan (1493) and after Juan's death to the court of Ludovico Sforza (Milan, 1499). From Milan, Oviedo moved to the court of Mantua in the service of Doña Isabel of Aragon, to the court of Fadrique of Naples, and back to Spain (1501) to the court of Ferdinand of Aragon.[21] He left the court circuit in 1507 when he moved to Madrid, married his first wife, Margarita de Vergara, and became a public scribe (1507). In 1513, after the death of his first wife, he obtained court clerkships in the Pedrarias expedition to Tierra Firme. Later he gained the royal position of supervisor of Tierra Firme's foundry.[22]

In the humanist courts of sixteenth-century Italy and Spain, Oviedo learned to use the wonder of the New World to fashion a space for himself. He played upon the novelty of the New World's nature to find royal support for his activities in the Indies. Both Ferdinand the Catholic and Charles V had obtained animals and plants from the New World. Charles V, for instance, owned a big cat from the New World (Oviedo called it a tiger) and was once presented with a pineapple that was sent "in its plant."[23] Charles's interest in natural history extended to his last days of life: at Yuste, where he retired to die, Charles asked his personal physicians about the "properties of fruits, herbs, and plants" at lunchtime.[24] Oviedo appealed to his patrons' interest in natural history. The study of nature was, among other things, a strategy for Oviedo to gain a better social position, just as it was for natural historians from Giuseppe Gabrieli (1494–1553) to Ulisse Aldrovandi (1522–1605).[25] When Oviedo returned to Spain in 1515, after his first trip to the Indies, he brought with him natural curiosities sent by Santo Domingo's royal treasurer, Miguel de Pasamonte, to King Ferdinand: parrots, sugar-breads, and samples of drumstick tree.[26]

After his return to Spain and the death of Ferdinand (January 23, 1516), Oviedo traveled to Flanders to gain the favor of the new monarch, Charles I (later Charles V of the Holy Roman Empire). He failed to obtain access to the king, however, and went back to Castile. When Charles arrived in Spain, he appointed new officials to the high ranks of the royal bureaucracy; and with them the system of bureaucratic patronage changed. Prior to Charles, the administration of the New World had been in the hands of the bishop Don Juan Rodríguez de Fonseca and his secretary, Lope Conchillos. After Charles came to power, Lope Conchillos was dismissed; and Fonseca was joined by Cardinal Adrian of Utrecht, the chancellor Jean le Sauvage, and a royal official named Zapata.[27]

Oviedo, who had been working with Lope Conchillos, had to wait before approaching the court again. His opportunity appeared when Bartolomé de las Casas began his campaign in favor of the Indians, when he presented a report before the Spanish Cortes in Barcelona in 1519.[28] Oviedo joined Fonseca in opposing las Casas; and that, together with his knowledge of the New World, helped him once again to obtain offices in the New World. He would remain this time from 1520 to 1523—in Panama, Santa María del Darién, and Hispaniola.

Back in Spain, Oviedo presented a report on the situation of Panama and Santa María del Darién to the newly created Council of Indies (1524). The council also asked him to give a report on the situation of the indigenous people. In 1526, as noted above, he published the *Sumario de la natural y general historia de las Indias.*

THE CHRONICLER OF NATURE

As mentioned earlier, in 1532 Oviedo was appointed "cronista de Indias" and was assigned the task of writing the social and natural history of the Indies. The Council of Indies consulted the king on May 27, 1532, about a proposal by Oviedo to write the natural history of the Indies:

> Describing in particular the characteristics of the lands and islands and the strange things that have been there or are there and the conditions of its dwellers and animals.[29]

The council explained that Oviedo had asked for a salary to visit with an assistant "those lands that he has not visited." It suggested that, for his "skill and experience," Oviedo be granted a salary so that he could write his history and include it in the history of Spain.[30] This was the first proposed natural history expedition to the Indies to collect empirical information about its natural products. Oviedo's expedition to the New World never took place; however, Charles V granted him a salary to write a history of "the things of the Indies."[31]

Having to remain in Spain, Oviedo had the problem of how to gather information about the places he had not yet visited. The crown dispatched decrees to royal officials in the Indies requesting information about the natural world. One, sent to the governor of Fernandina (Cuba), requested that information "about the island, its dwellers, and its conditions" be sent back to Oviedo.[32] Oviedo had to provide an annual copy of his writings to be added to the history of Spain. The informants from the New World were required to send signed information every time they were asked to do so, as promptly as possible.[33]

Oviedo's *Historia natural y general de las Indias* was the first result of his office. His natural history comprised two aspects of natural entities: the empirical description of single entities and their incorporation into a single framework for the understanding of this diversity. His work shared in the empirical culture that developed from the commodification of nature as well as from the bureaucratic interest in obtaining a complete description of nature.

Oviedo wanted his book of nature to be judged not by its literary merits but rather by "the novelty of what I want to say, which is what had moved me to write it."[34] He may not have discovered the nature of the New World, as one scholar put it, but he was the first to try to provide a global image of it.[35] In the 1570s the royal bureaucracy would try to implement a collective version of this project through the *Relaciones geográficas* and the expeditions of Jaime Juan and Dr. Francisco Hernández.

Oviedo's book of nature was written in the language of experience: sensory information, measures, and tests. Some natural products were difficult to describe. In these cases, it was necessary to draw an illustration to understand "by means of the vision what the language misses."[36] Oviedo provided few illustrations, and all of them alluded to Indian devices: a hammock, sticks for making fire, and the tree for building canoes (see Figures 6, 7, and 8).

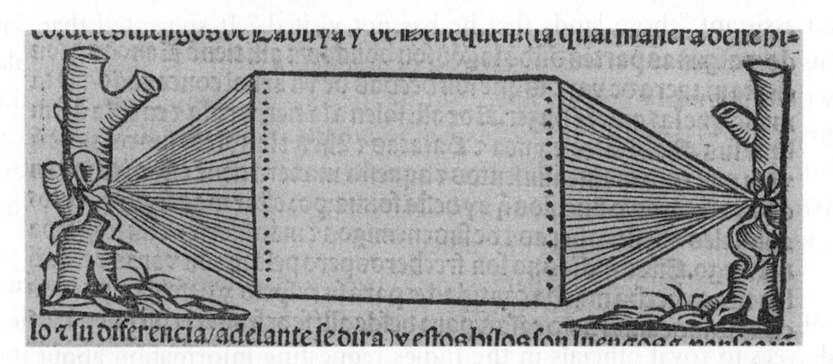

6. Hammock in Gonzalo Fernández de Oviedo y Valdés, Oviedo dela natural hystoria delas Indias *(Toledo, 1526). Reproduced courtesy of the John Carter Brown Library at Brown University.*

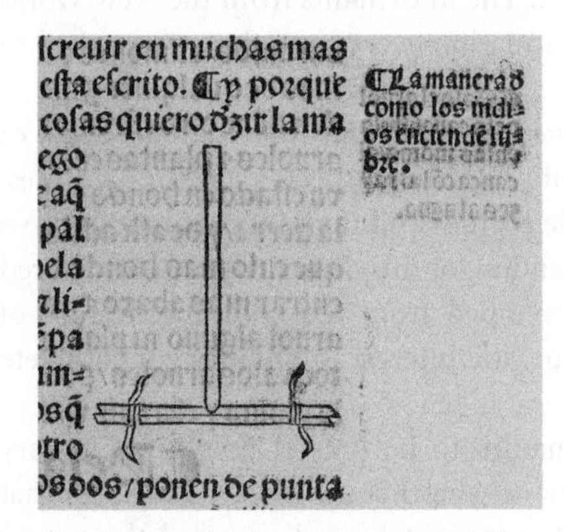

7. Sticks used by Native Americans to make fire in Gonzalo Fernández de Oviedo y Valdés, Oviedo dela natural hystoria delas Indias *(Toledo, 1526). Reproduced courtesy of the John Carter Brown Library at Brown University.*

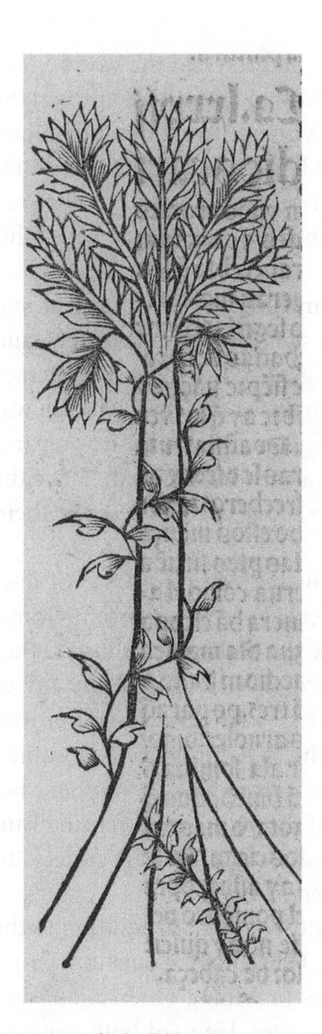

8. Tree for building canoes in Gonzalo Fernández de Oviedo y Valdés, Oviedo dela natural hystoria delas Indias *(Toledo, 1526). Reproduced courtesy of the John Carter Brown Library at Brown University.*

According to Oviedo, books on nature were more accurate and more authentic if the writer had personally seen and understood the works of nature, without any other interest save a natural desire to know them.[37] How true to things of nature such books were depended upon the "courteous understanding of the man who had traveled the world." The expert had the authority to provide and understand information about the world. Pliny, according to Oviedo, might have read "two thousand million books," but "I accumulated all that I wrote here from two thousand million works and scarcities and dangers in twenty-two years."[38] Oviedo's personal experience made the text true; and, "in accordance with" this truth, the text avoided "contradiction."[39]

The primary purpose of Oviedo's natural history was practical and utilitarian. He attempted to identify the uses of plants, animals, trees, and fish for human purposes, as had been the goal of the royal bureaucracy ever since the first trading contracts. Oviedo wrote that the guaiacum was good against syphilis and provided the indigenous recipe for its preparation. He explained that hog plum (*hobo*) was bad for the teeth. Since the guaiacum and hog plum were new realities, he needed to describe them thoroughly.[40]

Oviedo also performed experiments (in the restricted sense of tests or inquiries). In one case, he "wanted to see if mute dogs [from Nicaragua], removed from their land, would bark in another; and thus . . . [he] took a little dog of those from the province of Nicaragua to the city of Panama . . . and there it remained also mute."[41] This experiment proved that Panama and Nicaragua were part of the same province. He wanted to take this dog to Spain, but someone stole it from him. In another case, Oviedo marveled at the size and hardness of the anthill of a very small ant in Tierra Firme. These anthills were as hard as rock. He gave orders "to demolish and tear them down" to study them.[42]

In other cases, Oviedo kept animals for observation. In addition to the mute dog, he observed sloths and iguanas. He sent an iguana in a barrel full of soil to an Italian friend.[43] As well as providing observations and empirical information, Oviedo measured nature. He used both conventional measures (yardsticks) and body measures to add dimension and quantity to his descriptions of the New World. He measured a ceiba tree "with a string of hemp, and it had at its base a perimeter of 33 yardsticks or 132 palms."[44] A coconut was as big as a "closed hand"; some spiders were as big as an "open hand."[45] The information contained in this natural history could not be "learned in Salamanca, Bologna, or Paris," but only through experience.[46]

Oviedo used all of his senses to describe nature. The alligator smelled like musk; the meat of the armadillo had a very good flavor; the soursop was a beautiful tree. Things became the sum of their shapes, colors, smells, flavors, weights, and uses, according to the perceptions available to the human sensory system. Against vain or fantastic speculation, Oviedo relied on his senses: he experienced nature.

FRAMING PERSONAL EXPERIENCE

Oviedo's book of nature, however, was not entirely a product of his own experience. Personal experience makes sense only within a larger theoretical framework, and Oviedo found that framework in Pliny's writings.[47] Oviedo followed Pliny in the organization of his natural history but "not in saying what he [Pliny] said, for in some places his authorities would be contended."[48] Oviedo was, instead, "an ocular witness" who had "experienced in . . . [his] person" what he wrote about; or else he had collected several reports, "not giving credit to only one witness, but to many."[49] These were the same methods used by the Casa.

Pliny's model allowed Oviedo to integrate the natural entities of the New World with those of the Old World in a universe of diverse and interconnected regions.

> It is true that, according to the marvels of the world, creatures have different characteristics according to the different provinces and constellations where they are nurtured. While here plants are harmful and poisonous, there they are inoffensive and useful. Birds that in one given province taste good, in another are not even eaten. Man is black in some places, in others white, and the one and the other are men. . . . All these things, and many others that could be said in favor of this, are very easy to prove and worth believing for all who have read about or gone around the world, and whose own eyes will have shown them the truth of what is said.[50]

In Oviedo's natural history, the Old World and the New World shared the same system of relations; the diversity of natural entities was merely the product of diverse provinces and constellations. Empirical information provided him with the knowledge to support this conception of the natural world, and Pliny's model proved useful to the extent that it helped Oviedo to explain his experiences and information. He also used Pliny to provide his natural history with a structure and to integrate it into a tradition that gave him access to particular communities of scholars and courtly patrons.

But Oviedo's was more than a mere natural history—he gave his work a religious and political dimension as well. Oviedo's natural history mirrored the greatness of creation and its creator. God had led the Spaniards to the New World, thus bringing it under the possession of the Castilian crown. The nature of the New World, therefore, testified to the greatness of God as well as to the richness of Charles V. In the context of the Protestant Reformation, Oviedo's natural history was proof that the true God was with Charles and that he was the richest and most powerful of the European monarchs. The empirical study of nature became a way of approaching God. Nature, for instance, produced insects to annoy humans and thus remind "them of the principal duty to which man was created; that is, to know his maker and to procure his means of salvation."[51]

Nature and God were independent but interrelated: nature (not God) produced insects, yet they were there to remind humans of God. Reading books about insects, animals, and plants helps humankind to "praise and better know the creator and cause of all" things.[52] This was an old argument, of course, but in the context of the New World it provided a strong support for the use of personal experience in the study of nature. By the seventeenth century, it was a common argument.

At the same time, to know the order created by God was to understand the human order. In this order, Charles V rightly received from God the riches of the New World in order to defend his empire and religion better:

> No matter for what end the readers want to listen to me, I want them to listen and learn from me in all the world how rich is this empire of the Indies that God reserved for the blessed Emperor [Charles V].[53]

Nature became a map for salvation as well as political and religious propaganda. It was that garden of God where the king strolled.

Although empirical observations provided new information about God's creations, observations did not explain every natural event. In those cases where Oviedo could not understand something, he did not try to give any explanation—he simply gave all the information he had and hoped the events would one day be explained either by the collaboration of other people or by more information. For instance, he did not understand why the tides were different north and south of Panama. He asked cosmographers, astrologers, pilots, and natives of Panama about this problem, and no one could give him a convincing answer.[54] But Oviedo's empirical conception of nature provided him with a framework to understand that knowledge was based on cumulative information:

> I have talked about all this [the difference in the tides north and south of Panama] with some people of letters. They have not satisfied me either because they do not understand it or because I have not been able to explain it, or because they have not seen it as I have seen it. I, at least, am satisfied remembering that . . . [God], who allows those admirable things to happen, knows how to create this and other wonders, wonders that the human understanding does not reach without a special grace. I have put this problem here as a direct witness; I have not deserved to find its solution yet, but, truly, I would be very happy to see it solved.[55]

If personal and collective knowledge could not provide an answer to a problem, time would do it. There was always an answer to all natural wonders—if somehow a problem was inaccessible to human understanding, God could provide the solution to it by special grace.[56] Oviedo extricated the study of nature from classical authorities and moved it to the realm of empirical information for the understanding of God and the improvement of human life. In the second half of the sixteenth century, the Jesuit José de Acosta and collectors in Seville would continue Oviedo's tradition.

TRIDENTINE SCIENCE

In 1521 the ship *Victoria* came back to Spain after the first circumnavigation of the globe. The magnitude of this achievement is difficult to grasp today. The first expedition around the world in the sixteenth century was perhaps the equivalent of the first expedition to the moon in our time. Years after the expedition, José de Acosta exclaimed:

> [W]ho will not believe that this fact shows that all the vast size of earth, no matter how big it is painted, is subject to the feet of one man, for he was able to measure it?[57]

The successful circumnavigation of earth meant that it was possible to measure and control the earth—and that the means of doing so was by the actions of experts such as pilots. It meant also that such personal experience and actions were a significant source of knowledge about the earth and a source of power over it.

Acosta set himself the task of integrating personal experience, as a source of knowledge and power, into a new theory about the universe and a theory of knowledge. By 1590, when Acosta published his *Historia natural y moral de las Indias,* he could reflect on some sixty years of writing about American nature:

> Although the New World is not new anymore but old, for there is already much written and said about it, I still think that this *Historia* could be taken for new because it combines history and philosophy.[58]

The first two books were a translation from Latin of the author's *Natura novi orbis* (1586). Book 7 was an extract (with numerous literal translations) from Father Juan de Tovar's *Segunda relación.*[59]

José de Acosta was born to a merchant family in Medina del Campo. In 1551 he entered the College of the Jesuits there and in 1554 made his first religious vows. In 1557 he traveled to Plascencia, Lisbon, Coimbra, Valladolid, and Segovia, where he founded a College of Jesuits in 1559. That same year, Acosta entered the Universidad de Alcalá de Henares to study theology, canon and civil law, natural history, and history. He took orders in 1562 and resided from then until 1565 in Rome. From 1567 to 1571 he taught theology, first in the College at Ocaña and later at Plascencia. Acosta was an Aristotelian by training and a humanist by inclination. He would couple these two aspects of his intellect with his experience in the New World to provide a theory of the universe and knowledge.

In 1568 and 1569 Acosta asked to be sent to the New World; and in 1571 he finally made it to Peru.[60] As part of his administrative and religious activities there, he completed several trips around Peru (Arequipa, Cuzco, Chuquisaca, Juli, La Paz or Chuquiabo, Potosí) from 1573 to 1578. In Chuquisaca, where the viceroy Toledo had called for him, Acosta had access to the writings of the lawyer, magistrate, and antiquarian Polo de Ondegardo (in particular, to his *Información acerca de la religión y gobierno de los Incas*), which would later prove influential for his *Historia.*[61] Polo de Ondegardo also invited Acosta to see the Inca mummies that he had found in his jurisdiction. In 1572 Acosta was appointed president of the College of Jesuits at Lima, and in 1573 he was appointed Provincial of the Jesuits in Peru. During his tenure there, he traveled around Peru, convoked the first Congregation of Jesuits in Lima, and supervised the foundation of new colleges at Arequipa and Potosí. In 1581 Acosta requested a license to go back to Spain. He returned to Spain, after spending 1587 in New Spain.

In New Spain, Acosta met the Jesuits Juan de Tovar and Alonso Sánchez. From Tovar, Acosta obtained descriptions of the Aztecs. Tovar had based his description on Fray Diego Durán's *Historia de las Indias de Nueva España.* From Sánchez, Acosta obtained reports and information about China and Japan. In 1587 he arrived in Sanlúcar de Barrameda, where his old teacher from Alcalá de Henares, the Jesuit Gil González de Dávila, was waiting for him.

González informed the General of the Jesuits, Claudio Aquaviva, that Acosta was healthy and that he had written some books and brought news from the New World:

> He brings notes about things of the Indians, strange and very pleasant and even of some benefit that he could use to improve his book *De natura novi orbis.* His knowledge about those parts is great; he has improved it with his stay in New Spain.[62]

Acosta's book on nature brought the nature of the New World to Europe.[63] He began with a discussion of the structure of the universe, its round heavens, and the globe of earth. He then discussed different opinions on the nature of the earth and the Tropics and his own experience in this region. Acosta reviewed specific problems, such as the movement of the celestial bodies, the antipodes, the origin of Native Americans, and the nature of the Tropics.[64] After this discussion, he dealt with the type of winds and waters and the characteristics of the land then, finally, with the natural history of the region, plants, animals, and metals.[65]

Acosta had a different aim than Oviedo. He was less concerned with describing natural entities and more interested in natural causes and principles. Acosta complained:

> Many authors have written diverse books and accounts about the New World and West Indies in which they report about the new and strange things discovered in those lands, and about the facts and actions of the Spaniards who have conquered and inhabited them. But until now, I have not seen any author who tries to find the causes and reason of those novelties and wonders of nature . . . nor I have found a book whose argument deals with the facts and history of the Indians and native population of the New Orbus.[66]

Classical sources could not account for these principles and causes. For Acosta, the New World continued to appear new for the Europeans because the New World's "things of nature" were

> outside the classical philosophy already received and discussed. This is the case of the region called Torrid. [It is] very humid, and in other parts very temperate. It rains when the sun is very close, and other similar events.[67]

Things of this nature were strange because they did not behave or appear according to the precepts of classical philosophy. The nature of the New World became an independent field of study. The same was true of the natives of the New World—they were strange not only because they lived outside the classical framework but also because nobody had "a close and uninterrupted relationship with the Indians" and thus tried to understand them.[68]

Acosta described the universe as a machine manufactured by God.[69] The New World was part of this machine and, of course, shared in its characteristics. Acosta was not by any means the first to integrate the New World and the Old World in this way. Jerónimo de Chaves in his *Tratado de la sphera* (1545) had already discussed the place of the New World within the *machina* of the universe.[70]

The notion of machine (Latin *machina*) was applied in the sixteenth century to artifacts composed of diverse elements. Thus, the crown granted patents to machines made of wheels and bars, like mills or water purifiers.[71] In the texts of Chaves and Acosta, the world machine was conceptualized in the same way—a machine is a composition of diverse parts organized toward the same end.[72] He compared the movement of the celestial bodies and the heavens with the movement of "spokes and the wheel."[73]

Fernández de Oviedo had placed the New World within the Pliny tradition of diversity; Acosta framed its nature as a machine, in the organic sense just discussed. This allowed him to look for causes in the New World despite the lack of classical markers in it. It also allowed him to link his study of nature with God.

Acosta began with the basic Christian understanding that God created the world. The machine of the universe was organized according to God's laws: all "created nature obeys promptly . . . [God's] hidden laws."[74] Through the empirical knowledge of particular entities or events humans could understand the secret causes or laws of the universe.[75] "In the natural and physical causes, one should not be asked for an infallible and mathematical rule, but rather for what is ordinary and common, for that is what constitutes a rule."[76] Thus, Acosta launched an empirical program for the study of nature based on particular entities and personal experience similar to the Casa's program.

Acosta placed the chain of creation, the particular entities of nature, within a familiar framework: "Always the inferior creature nourishes the superior one; and the less perfect is subordinated to the more perfect one."[77] Fertile soil provided nourishment to plants, plants to animals, and animals and plants to humans. In this sense, nature was a self-sustained system. Natural entities were interconnected in a hierarchy from the simplest to the highest. The secret causes explaining the facts of the universe were framed in relative and average terms rather than in mathematical or absolute terms.

Acosta's conception of the universe was closed to human changes. Humans could not transform the creations of God. If they tried to change nature, for instance, "knocking down" the "wall" in Panama that divided the Pacific from the Atlantic, "it would be fair to be afraid of God's punishment for trying to amend the works that the Maker, with great understanding and providence, ordered in the fabric of this Universe."[78] The desire to reorganize God's machine was to attempt to transform God's order. People could not alter this finished machine, but they could understand and measure it for their own salvation.

Once Acosta had situated the nature of the New World within a coherent theory of the universe, he moved to deal with the peculiarities of the natural world of the Indies. These peculiarities not only did not fit within the "classical philosophy already received and discussed" but also introduced innovations in that philosophy and in the structure of knowledge. This was the case with the shape of the world, as Acosta states at the opening of his discussion:

> The ancient [thinkers] were so far away from thinking that there were people in the New World that many of them did not want to believe that there was land on this part. What is even more extraordinary, there were those who denied that there was this sky that we see here.[79]

Thinkers such as Saint John Chrysostom, Theodosius, Theophilus, and Saint Augustine, said Acosta, "imagined the fabric of the world in the manner of a house": the heaven on top and nothing, or chaos, underneath the world.[80] This conception of the world implied the negation, of course, of the circular movement of the heavens. In contradiction to their view, experience had shown that the earth was a globe encircled by the heavens. Even more, experience provided information about the existence of lands and people in a region where, according to Aristotle and other classical sources, none such existed. Acosta commented:

> There were so many and admirable properties [in the Torrid Zone] that with good reason [these properties] awaken and enliven the understanding to inquire into their causes, guiding us toward true reason and certain experience rather than the doctrine of the classical philosophers.[81]

Acosta also excused other authorities, in this case the doctors of the church, on the grounds that they did not really study natural sciences but, rather, religious matters. In contrast, the "wise people and vain philosophers of this century" were reprehensible, for although

> these unfortunate people know the being and order of the creatures and the course and movement of the heavens, they have not come to know the Creator and Maker of all this.[82]

If for the doctors of the church the study of God was more important than the study of nature, for Acosta the study of nature was a means of approaching God. He was following the tradition already established by Oviedo and others who had studied the nature of the New World in order to reach God.

Acosta problematized experience in a way that Oviedo did not. The Spanish bureaucrats had already placed two conditions on the gathering of empirical information. First, information had to come from multiple reports, as was the case with the juntas at the Casa de la Contratación and the informants for the questionnaires of 1577. Second, information had to come from experts or people with experience, as was the case with pilots or residents who provided information about unknown areas. Oviedo sought to use only information supported by multiple sources or by his own experience. Acosta continued this empirical tradition but placed less emphasis on its specific mechanisms; he provided a theory for restructuring the place of empirical information within the hierarchy of cognitive functions.

The problem for Acosta involved assembling pieces of empirical information into a whole. He proposed the case of the shape of the earth as an example of the difficulty of gaining knowledge. Experience demonstrated that the earth was a globe. If a ship left Seville toward the West and navigated continually in that direction, it would return to Seville from the East.

The problem, as Acosta put it, was that the earth's shape had not been seen—rather, it had been imagined, thanks to the experience gained. For Acosta imagination and experience were two elements in the process of knowledge. Experience furnished the imagination with information, and then the imagination reorganized the information. But how?

The problem of knowledge is its universal character. Imagination and empirical information are both localized. The imagination, argued Acosta,

> is subject to a [particular] time and place, and it does not perceive the same time and place universally, but locally. Thus when the imagination is lifted to consider things that exceed and surpass the known time and place, it fails. If reason does not maintain and lift it, [the imagination] cannot stand up.[83]

According to Acosta's theory, imagination needed the guidance of reason to organize empirical information in universal terms. Otherwise, it could fancy anything, as the doctors of the church and Aristotle sometimes did.[84]

Reason lifted the products of the imagination to the rank of knowledge. But reason also needed helpers to move it in the appropriate direction. Only the imagination could assemble empirical information about the world in order to reach the conclusion that the "heavens are round" and the "earth is in the middle."[85] Reason checked the imagination before it conjured up "men who resided on the other side of earth walking with their feet up and head down." Reason dictated that heaven, everywhere, "is up," and earth, everywhere, "is down."[86]

Acosta suggested another limit to the imagination—religious teachings. In this case, problems such as the creation *ex nihilo* of the universe could only be framed within religious teachings. These teachings, for Acosta, were rational. Thus, he argued that "reason clearly shows that there was neither time before the creation of movement, whose measure is time, nor place before [the creation of] the universe, which encloses all places."[87] For Acosta, reason partook in the "divine light" through which it judged "the same images and interior forms that are given to our understanding" by the imagination and empirical information. He concluded that those who did not know about or who doubted this divine light in human reason did not know about their humanity.[88] Imagination's anchor was reason; reason's anchor was faith.[89] Faith was articulated around a universal concept of the human being, with the implicit assumption that knowledge was universal.

Since humankind shared in God's "light," people could attempt to grasp the laws that regulate the world's machine. This enlightened reason put limits to human knowledge at two levels. First, reason's grounds were given by faith and religious coherence. Second, human reason was limited in its capabilities by its own humanity. In limiting empirical reason by faith, Acosta had to address the relationship between the Bible and natural history. Acosta's conception of knowledge forced him to move, in some cases, from a literal reading of the Bible to a symbolic one. Thus, a literal reading of Psalm 135:6 (that earth was established above water) made sense to Acosta.[90] But he interpreted Job 26:7 (that the earth was placed in the universe over nothing) as meaning that the earth was placed in the middle of air. It did not fall from there because it had "secure foundations from its natural stability."[91]

In another case, Acosta dismissed the idea, held by some authors, that the New World corresponded to Ophir and Tarsis (the lands whence King Solomon obtained gold, silver, and ivory) by explaining, among other things, that there were no elephants in Peru.[92] He also contested other identifications of Ophir and Tarsis and concluded that these terms meant either "the immense sea or far away and distant regions."[93]

Acosta's decision to use religious texts to complement his empirical knowledge was first of all, of course, a matter of faith, but it was also a methodological solution to the problem of the limits of empirical knowledge. In his theory, empirical information provided clues for interpreting the Bible. Experience not only furnished the raw material for knowledge; it also constituted a key element in fulfilling human salvation. With the appearance of Acosta's book of nature, the sixteenth-century empirical tradition of understanding the New World was finally situated within a theory of knowledge that was subordinated to a theory of the universe.

Acosta's theory provided the Jesuits with the elements needed to confront the scientific empirical production of other European groups at this time of rapid intellectual change. Acosta, together with Oviedo, created an intellectual tradition that constituted yet another version of modern science, as it was defined at the end of the sixteenth century. At the limits of this knowledge stood God. After all, the purpose of empirical knowledge was either to know God more fully or to provide a better government. This, too, in the end, was a moral imperative.

COLLECTING NATURE

Initially, curiosity seekers had little access to New World marvels. During the first half of the sixteenth century, few animals and plants from the New World entered the Old World. Collectors who did manage to import marvels were mostly connected to the court, humanist circles, and explorers. As mentioned earlier, the humanist Peter Martyr obtained a royal decree that ordered shipmasters to bring animals and plants such as parrots, "turkeys from Tierra Firme," "other strange birds," fruits, iguanas, chiles, cinnamon, roots, blue stones, amber, or "anything" that the officials from Hispaniola sent him.[94] Martyr wanted to show these marvels at court. In 1547 Oviedo brought a *cozumatle* (a type of cat, according to him) to Madrid and presented it to one of his relatives from Asturias.[95] From Hispaniola, Oviedo shipped an iguana in a barrel full of soil (as the iguana's only nourishment) to his Venetian friend Joan Baptista.[96]

Explorers also knew that marvels appealed to the goodwill of courtly patrons. In 1526 Charles V received an *ochi* or "tiger," from New Spain.[97] In 1528 Sebastian Cabot sent him three llamas from Peru.[98] In the 1530s a conquistador sent to the queen a marvelous pet (another type of cat) that belonged to his Inca wife, but it died during the voyage.[99]

Common people (sailors, soldiers, farmers, and priests) also trafficked in New World curiosities. Chiles, according to Oviedo, were quite popular in Spain and Italy. *Batatas* (a type of sweet potato) were less successful in making the trip to Europe: they often rotted. In 1530 corn was already growing in Avila, Spain.[100] Many different groups were involved in the traffic in commodities and curiosities, usually for their own benefit. These social groups thus came to understand nature as a collection of manipulable entities—this understanding was part of the culture emerging in the Atlantic world.

The second half of the sixteenth century witnessed an increased interest in curiosities, legitimizing an empirical approach to nature; the result was new gardens, museums, and the first natural-history voyage of exploration to the New World. During this period, the trafficking in curiosities became more systematic. In 1554 Philip II, a lover of gardens, ordered his officials in Hispaniola to send "all the seeds and plants found" on the island. They sent him some "plants of pineapples" and other plants.[101] Philip also organized the first natural-history expedition to the New World. In 1570 he sent Dr. Francisco Hernández to collect samples and information about medicines.[102] In 1574 Dr. Hernández sent some "seeds, plants, and other natural things" to the king.[103] In 1578 someone else sent two trees from Chile, one of them a balsam tree, for the king's garden.[104] New World curiosities and commodities had put the empirical approach at the center of the study of nature.[105]

Furthermore, New World curiosities attracted to the study of nature a diverse band of professionals who shared common knowledge.[106] Before long, a group of collectors emerged in Seville.[107] Seville, however, was not the only center of collecting: the surgeon Pedro Arias de Benavides, from Toro, Castile, established an exchange of seeds and plants with the entrepreneur Bernardino del Castillo in Mexico. But the Seville group was large, and its activities significant. It consisted of professionals (physicians, cosmographers, surgeons, royal officials), some of them engaged in mercantile activities, who established gardens and museums to house and grow New World products. Nicolás Monardes (ca. 1493–1589), a physician and entrepreneur, is the best known of this group. His father, Niculoso de Monardes, was a Genoese bookseller established in Seville. Monardes obtained his bachelor's degree in art and philosophy in 1530 and a bachelor's in medicine in 1533, both at the Universidad Complutense.[108] In 1547 Monardes received the *licenciatura* and physician's degree from the Colegio-Universidad de Santa María de Jesús de Sevilla.[109] His medical practice in Seville was very successful. He worked with Doctor Garciperez Morales and married Morales's daughter Catalina in 1537.[110]

Monardes presumably was introduced to the curiosities of the New World through Dr. Morales, who had written a treatise on Villasante's balsam. Monardes was very successful in his commercial dealings with the Indies and took advantage of his partners to obtain curiosities and commodities. His commercial partner, Juan Núñez de Herrera, sent seeds and plants for Monardes's garden in Seville. Monardes grew many New World plants and herbs in his garden and wrote a *Historia* based on his work there.[111] In his plants and herbs he saw not only marvels but also the possibility of "incredible wealth."[112]

Monardes was not the only one collecting curiosities from the New World, establishing networks of information, and incorporating empirical practices in the study of nature. Simón Tovar, another physician, had a well-known garden from which he sent samples and information to foreign correspondents. He elaborated an annual catalog of his plants—the first in Europe, the *Index Horti Tovarici*—and distributed it among botanists in Europe.[113] The well-connected botanist Carolus Clusius not only received Tovar's catalog but also visited his garden when he traveled in Spain between 1564 and 1565. When Tovar died, Philip II expressed interest in buying the garden.[114] Clusius also met the physician Juan de Castañeda, who collected plants from the New World and worked at the hospital of the "Flemish nation" in Seville. Once in Leiden, Clusius would receive information and samples from Castañeda's garden.[115] The gentleman Gonzalo Argote de

Molina had a "famous museum" in Seville housing books, horses, arms, animal heads, paintings, coins, stones, and "animals and birds, and other curious things brought from both the East Indies as well as the West Indies and other parts of the world."[116] Monardes visited this museum to see and study some of the animals he was describing in his history (e.g., the armadillo; see Figure 9).[117] In 1570 Philip II even visited the Argote museum, in disguise, in order to experience its wonder as an ordinary visitor.[118]

Curiosities appealed not only to physicians and botanists but also to religious officials and cosmographers. Monardes mentions that the archdeacon of Niebla received a *ciervicabra* (a goat-deer) from "very far, via Africa."[119] The archdeacon might have been collecting curiosities from the New World as well. The bishop of Cartagena brought from Tierra Firme "the fruit of the tree from which dragon's blood is obtained" and seems to have brought other herbs and medicines as well.[120] In 1525 Fray Juan Caro (working for the Portuguese crown in Cochin, India) sent his relative in Seville, Dr. Porras, a "parrot from China, bright red and with many other colors, that speaks quite well." Unfortunately, he did not mention what language the parrot spoke.[121] This movement of curiosities not only helped to establish networks of information among royal officials, pilots, prelates, physicians, and adventurers but also helped to promote a common understanding of nature.

Two cosmographers from the Casa also collected curiosities. Jerónimo de Chaves had "a piece of a whale," "three sticks from the New World [effective] for [healing] the loin," "an ostrich egg," "a turtle shell and a snail," "a box full of *Mechoacán*," "a box full of shells," "three Turkish bows with a quiver full of arrows," "three bows from England and from indigenous people," "a glass jar with preserved artichokes from the Indies," and "a vial with balsam from the Indies."[122] The cosmographer Rodrigo Zamorano had a larger collection of animals (armadillos, snails, shellfish) in his house. He had gathered his specimens with the help of pilots and shipmasters: "Each shipmaster that goes [to the Indies] happily brings him some new or extraordinary things, and thus he had the walls of his house full of shells, fish, and animals worth seeing."[123]

Collections of American natural entities served several purposes. Tovar's and Monardes's gardens were places for research. Monardes used his garden to study new plants and to conduct research on new medicines. His *Historia* became very popular not only among physicians but among explorers as well—both a Spanish soldier in Peru and an English merchant in Virginia used it to search for commodities in the New World (see Figure 10).[124] Collections also promoted the reputation and

EL ARMADILLO.

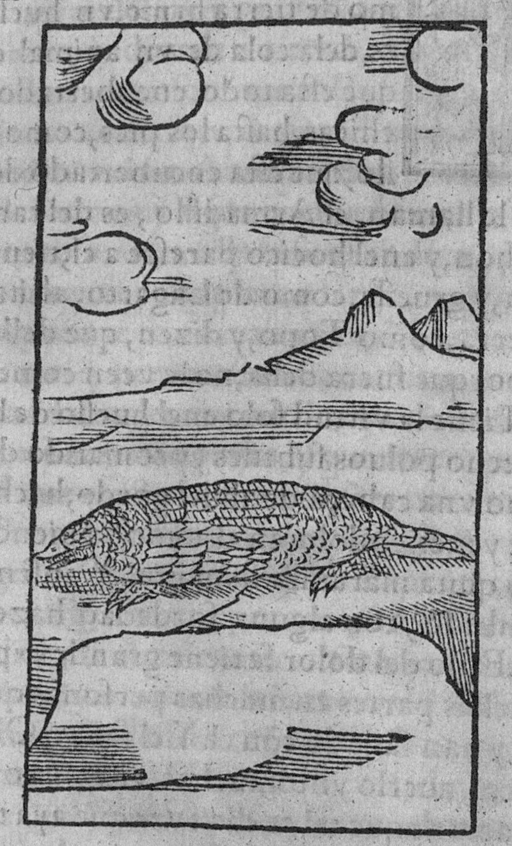

ESTE animal ſaque de otro natural, que eſta enel Muſeo de Gõçalo de Molina, vn cauallero deſta ciudad, enel qual ay mucha cãtidad de libros de varia lection, y muchos generos de animales y aues, y otras coſas curioſas, traydas aßi de la India Orientol, como Occidental, y de otras partes del mundo, y gran copia de monedas y piedras antiguas, y diferencias de armas, que con gran curioſidad y con generoſo animo ha allegado.

9. Armadillo in Nicolás Bautista Monardes, Primera y Segunda y Tercera partes de la Historia Medicinal de las cosas que se traen de nuestras Indias Occidentales que sirven en Medicina *(Sevilla, 1574). Reproduced courtesy of the John Carter Brown Library at Brown University.*

10. Sassafras in Nicolás Bautista Monardes, Primera y Segunda y Tercera partes de la Historia Medicinal de las cosas que se traen de nuestras Indias Occidentales que sirven en Medicina *(Sevilla, 1574). Reproduced courtesy of the John Carter Brown Library at Brown University.*

careers of their proprietors: Philip II, as mentioned earlier, visited Argote de Molina's famous museum incognito. Monardes was well known outside of Spain; foreign botanists such as Clusius visited Tovar's garden.[125] Spanish collectors had an advantage over other European collectors such as the Italian botanist Ulisse Aldrovandi, who did not have regular access to New World products and thus rejected them as impractical for collecting.[126]

Although the trafficking in curiosities often involved professional connections, lay readers of books on curiosities also joined the network of communication and trade. Monardes and Tovar received most of their samples from their mercantile partners in the New World. Zamorano and Chaves had constant dealings with shipmasters and pilots. But Monardes occasionally received curiosities from laypeople who had read his work. A certain Pedro de Osma—a soldier who had read Monardes's *History*—sent him a collection of curiosities: a box with twelve bezoar stones that he had found in an "animal that lives in the mountains of this land [Peru], and that looks very much like a ram"; a box with beans from Peru and instructions for cultivation; an herb, also from Peru, efficient in healing rheum and sore throat; and fruit from a tree found only in Peru that "is of great benefit." Apart from these curiosities, Osma sent information about other medicines and the ways indigenous people used them.[127]

Like Villasante, Osma used indigenous informants to gain access to the New World's curiosities and sent that information to Europe. Carolus Clusius translated Osma's letter into Latin and included it, together with excerpts from Monardes's *Historia,* in his *Exoticorum libri decem.*[128] Thus, curiosities helped in the establishment of networks of information encompassing indigenous people and explorers in Peru, doctors and collectors in Seville, and naturalists and physicians in Netherlands, France, Italy, and England.

The relationship between the physician Juan de Castañeda and Carolus Clusius is an example of how these networks functioned. Around 1600 Clusius wrote to Castañeda, apparently requesting information and samples of plants from the New World. Perhaps Clusius was trying to obtain new information for his coming book *Rariorum plantorum historia* (published in 1601). In response to this letter, Castañeda sent him the samples requested, plus a list of herbs from the New World then available in Seville: avocados, maguey, "the herb *tuatua* [*Jatropha gossypiifolia*], efficient against poison," "tomatoes like hearts," cassava, *chicama,* guava, plantain, *mamey,* maize, onions from Caracas, *ditamo* from Honduras, sapodilla, *curú, campanillas azules,* jalap or *mechoacán* (*Ipomoea purga*), *cabuya, capinos, anones, mamones, yerba viva,* "the one that [José de] Acosta calls *yerba mimosa,*" and many others. Castañeda concluded: "I know well that you would appreciate more of these herbs since they come from the Indies, and some . . . just recently have arrived." He was ready to send more plants upon Clusius's request.[129] This was the beginning of a correspondence that would last at least until 1604.[130] During this time, Castañeda sent Clusius information about curiosities (plants, animals, fish) arriving from the New World, samples of plants, and drawings.

Soon after his first letter, Clusius sent a second letter requesting more samples and information. This time Castañeda provided not only the plants that Clusius requested but also some new ones that had just arrived from the New World. The private commerce in plants from the New World was already very intense by the beginning of the seventeenth century. Clusius requested, in particular, *costo*—an ancient simple essential for theriac (a mixture of drugs thought to be an antidote for poison)—but Castañeda could not find it.[131] Castañeda was very interested in sending information and samples before Clusius published "his book." He also offered to make drawings of Zamorano's curiosities for Clusius's next project.

Perhaps prompted by his friend or by the collecting culture of Seville, Castañeda began his own garden. He does not explain how he obtained his samples, but he probably paid shipmasters, pilots, and merchants to bring them. By 1603 Castañeda claimed "to have . . . more than 30,000 rootstalks." Unfortunately, heavy rains destroyed his gardens on December 20, 1603.[132]

Oviedo, Acosta, Monardes, and Castañeda became part of networks for the circulation of information and commodities. These networks brought together groups from Mexico and Peru to Seville and Leiden, and each group brought a different level of experience and expertise about the New World. From the very beginning of the expansion into the New World, natural historians, physicians, entrepreneurs, and collectors sought new information and samples. They placed their own experience, reports, and collections at the center of new epistemological approaches to nature. From these activities and networks emerged the empiricism that characterized the new scientific practices of the sixteenth and seventeenth centuries.

Conclusions

THE POLITICS OF KNOWLEDGE

Spain's encounter with the New World launched Europe into the first imperial age of the modern world. By the sixteenth century, ships, charts, guns, Genesis, and the New Testament had intertwined in a Christian ideology of domination. Yet technology and God were not enough to establish an empire: an empire was and is, above all, the product of communication and information. Knowledge forms the lifeblood of any empire. The activities of artisans, merchants, royal officials, and entrepreneurs in America constituted the early Scientific Revolution. This means, first, that the Spaniards validated personal experience as a source of knowledge. As opposed to the textually based scholastic and humanist traditions, in the empirical tradition emerging in America personal experience provided information about particular events, curiosities, and commodities that became the basis for knowledge about nature. Second, these activities provided information that contradicted the authority of the classical tradition and its limitations for sixteenth-century Europeans living in America, for instance. Pliny did not have information about avocados or tomatoes. Most classical authorities were mistaken about the Torrid Zone and the antipodes.

The information collected in the New World helped not only to criticize classical tradition but to undermine it. It is not a surprise to have a fully new science by the mid-seventeenth century dismissing classical authorities and articulating personal experience in reports about nature. The early phase of this process began with the activities of artisans, entrepreneurs, royal officials, and merchants (and their interrelations) in the

New World and in the institutionalization of these practices at the Casa de la Contratación and Council of Indies.

Empirical practices found their place in institutions at the service of the state—perhaps the main difference from Portugal, which also developed an empirical culture in its books but did not institutionalize empirical practices to collect information, as did the Spaniards in the Casa. These institutions developed a set of rules for gathering information, schools for professionals, and books and reports. Spaniards in the New World eagerly sought information about routes and rivers necessary for establishing and improving communications; information about medicines, water supplies, and woods, indispensable for surviving and living in the New World; information about mines, mining techniques, and commodities, crucial for developing the economy of the Atlantic World; and information about people, their culture, social organization, religion, and military capabilities, which was essential for controlling them. Knowledge about navigation, ethnography, natural history, cosmography, and medicine became indispensable for establishing the Spanish empire, Spanish American communities, and networks of trade and communication between Spain and the New World.

These empirical and collective practices emerged from the relationship between private initiative and royal support. In most cases, the crown appropriated private initiatives and later launched them as state initiatives. Royal support was a condition for institutionalizing empirical practices. Without royal support, these institutions could not emerge or expand. The Casa de la Contratación, originally founded for the regulation of trade with the Atlantic islands, expanded its functions to include a veritable chamber of knowledge regarding navigation, cosmography, and geography. At the Casa, the crown institutionalized practices for gathering and organizing information, educational activities for the dissemination of new information, and certification mechanisms for verifying the quality of expert personnel. All of these activities relied on empirical practices and collective procedures. These empirical and collective practices created a community of experts relevant to the empire.

Thus, sixteenth-century political and economic interests redefined the aims of and tools for knowledge production. In the face of the New World, classical texts offered little useful information about the New World's nature. Certainly, classical texts and authors framed Spanish understandings of nature; yet in the process of collecting information the Spanish bureaucracy and imperial agents were not concerned with natural philosophy. Their epistemological attitude was practical and was guided by an interest in searching nature for commodities, curiosities, and information. This attitude is one of the key elements in the establishment of epistemological practices of this period. The process of moving outside classical frameworks of knowledge into modern ones was fostered greatly by the conquest, settlement, and incorporation of the New World into the European sphere of influence.

The New World helped to create a community of experts with the experience to describe the New World. These experts established networks of communication between various groups with different types of knowledge. The rapid development of mining techniques, for instance, shows how well connected and informed miners were about each other's improvements over instruments and technologies. In some cases, miners and adventurers relied on their counterparts in the Old World for finding new ways to exploit minerals in the New World. Thus, Juan Tetzel traveled to Spain and Germany with copper samples from Cuba to find ways to exploit it; Bartolomé de Medina discussed amalgamation techniques in his Seville tailor shop with a mysterious German master; and Juan Alemán received a report from Germany with a new method to exploit poor silver ores.

Yet miners were not alone in establishing communities of experts in the New World; natural historians, sailors, physicians, and merchants created their own communities of communication. Thus, the soldier Pedro de Osma began searching for curiosities such as bezoars among Andean animals after reading Monardes's book. Moreover, he sent Monardes curiosities and reports. Monardes published Osma's reports, and Carolus Clusius translated them in his *Exoticorum libri decem.* Pilots brought curiosities and commodities to cosmographers and physicians in Seville. In turn, these physicians and cosmographers sent information and samples to natural historians in other European cities. Oviedo and José de Acosta managed to obtain reports from witnesses and to incorporate them in their own natural histories. From these loosely interconnected communities emerged empirical practices for studying and discovering the secrets of the New World. How successful was this empirical program developed in Spain during the sixteenth century, in particular the programs established at the Casa de la Contratación and the Council of Indies for gathering information?

There are several answers to this question. From one point of view, it is possible to argue that the program succeeded to the extent that the Spanish crown was able to negotiate the control of America with the New World's residents up to the late eighteenth century. The program also succeeded to the extent that the people supporting it produced a considerable body of scientific and technical literature along with maps and instruments that specifically addressed scientific and technical problems raised by the control of the Atlantic world—this became the model, without doubt, for the English and Dutch empires. The sixteenth-century literature on cosmography, navigation, and natural history contributed to Spain's imperial development and influenced emerging European scientific practices in general. This is particularly clear in the publication and translation of Spanish works in the rest of Europe during the sixteenth century.

Martín Fernández de Enciso, for example, published in 1519 the first geographical treatise that discussed the Old and New Worlds together (Spanish editions in 1519, two in 1530, and in 1546). The geographical part of his book was translated into English.[1] In 1545 Pedro de Medina published his *Arte de navegar.*[2] In this book, Medina explained what pilots needed to know to obtain their certification as pilots from the Casa. His book was translated into French (fifteen editions between 1554 and 1633), Dutch (five editions between 1580 to 1598), Italian (three editions in 1554, 1555, and 1609), and English (two editions). In 1552 Medina published his *Regimiento de navegación,* a brief edition of his *Arte* without the theoretical aspects.[3] A year before, Martín Cortés de Albacar (d. 1582), who had been teaching cosmography and the art of navigation to pilots in Cádiz since 1530, published his *Breve compendio de la sphera y de la arte de navegar.*[4] He discussed in a very clear way matters of navigation and cosmography and dealt with problems such as the magnetic declination, for which he provided the hypothesis of the celestial magnetic pole. His treatise was translated into English (ten editions between 1561 and 1630). Cortés's magnetic theory was criticized by Robert Norman in his *The Newe Atractive* (1581) and by William Gilbert in his *De magnete* (1600). Cortés's hypothesis was discussed by Edmund Halley, Leonhard Euler, Pierre Charles Lemonnier, Georges Buffon, and Jérôme Lalande. Martin Frobisher and Francis Drake enjoyed and used Cortés's book.[5] His book became the navigational manual of Europeans through the early seventeenth century.

In 1545, the same year that Medina published his *Arte de navegar,* the cosmographer and collector Jerónimo de Chaves published his *Tractado de la sphera.* Three years later, he published a *Chronographia o Repertorio de los tiempos,* which would become a best seller, with thirteen editions by 1588. In this book he discussed calendars, the order of heavens and planets, astronomy, and astrology. Abraham Ortelius published some maps by Chaves in his *Theatrum orbis terrarum.*[6] In 1581 another cosmographer and collector, Rodrigo Zamorano, published his *Compendio del arte de navegar* (Seville), a manual of instruction for pilots (at that time he was the Casa's lecturer of cosmography). Zamorano used the mathematics of Copernicus to correct the tables of solar declination. There were four editions of this work in Seville before the end of the century, and it was translated into Dutch and English.[7] Finally, José de Acosta's *Historia natural y moral de las Indias* (1590) was translated into French (1598, 1600, 1606, 1616, 1617); Italian (1596); Dutch (1598, 1624); German (selections in 1598, 1600, 1605); and English (1604).

These references suggest not only the great interest among Europeans in the work of the Spaniards but also the interest in Spain to learn more about the New World. As I have argued, the key element in these books is the use of personal experience and the mechanisms to collect this information and transform it into knowledge. These translations offer not only practical knowledge but also (and perhaps more important in the long run) arguments for the validation of personal experience and personal reports in the production of knowledge.

This is not by any means a complete list of what these and other authors published during the sixteenth century. It indicates the types of books that were published in sixteenth-century Spain and the great interest of other European countries in these books. Europeans were not only interested in Spanish books; they were also interested in Spanish institutions. The Casa de la Contratación, for instance, became the model for a (planned) similar English institution. John Borough (d. 1570), who sailed in the service of John Dudley, Lord Lisle, and Stephen Borough (1525–1584), who served under Richard Chancellor as "chief pilot after the Spanish style," employed Spanish and Portuguese instruments, books, and charts. Stephen Borough visited the Casa in the mid-sixteenth century and afterward persuaded the Muscovy Company to translate Martín Cortés's *El arte de navegar* (Richard Eden translated it in 1561).[8] Richard Hakluyt, whose translations and collection of many Spanish texts on the New World made him a leading authority on Spain's colonial activities, praised the Casa de la Contratación and its navigational activities.[9] The Atlantic world, through Spain, contributed greatly to the development of empirical practices during the sixteenth and seventeenth centuries. The connection with the New World helps to explain why empirical knowledge became the key element in the production of knowledge during the sixteenth and seventeenth centuries. By the early seventeenth century, when Galileo was providing proof to support the Copernican system, there was already a century of personal reports undermining classical authorities and their geographical ideas—Ptolemy's ideas.

By the late sixteenth century, Spanish American communities had developed their own interests and activities in regard to the study and understanding of the New World. Spain and America were already moving along different paths regarding natural knowledge.

In Spain, the late-sixteenth century financial problems of the crown greatly affected the development of natural knowledge, cosmography, and technology outside royal institutions. The Thirty Years' War and the war against the United Provinces (the Netherlands) drained Spain of material resources. By 1648 Spain had lost the war in Europe, and most of the intellectual debate in Spain was about political and social issues.[10] Regarding the New World, activities such as natural history, cosmography, and technology were no longer necessary as they had been during the sixteenth century for the establishment of economic activities and political jurisdictions. Once again, the economic and political context shaped the development of these activities.

Yet by the early seventeenth century American communities had already developed their own identities and particular interests. For instance, Juan de Cárdenas (1563–after 1591)—who moved to Mexico when he was fifteen and studied at the University of Mexico—wrote a natural history on American issues: the influence of climate and nature on the Spanish character, the cultural significance of some American foods such as chocolate, the connection between peyote and hallucinations. Alonso Barba (1569–after 1640), who had moved to Peru when he was about eighteen years old, wrote about every aspect of the amalgamation process and its instruments. Both authors, who were contemporary and received their formation in America, situated their work within American realities and developed themes closely connected to those realities.[11]

As the Spanish crown had used natural history and cosmography to create its Atlantic empire, the new emerging American communities would use natural history and cosmography to create their own territorial and cultural identities. The political uses of those activities with their empirical and collective emphasis are structurally linked to the formation of modern states.

In the process of establishing a long-distance empire, Spain provides a paradigmatic case to explore the relations between politics and knowledge. The establishment of long-distance empires helps to explain the increased emphasis on empirical and collaborative practices that emerged during the sixteenth and seventeenth centuries in Europe and America. Such practices became central to a general definition of science, not just in Spain but throughout Europe by the mid-seventeenth century, as manifested in the several scientific academies that emerged in Europe at the time. As in Spain, European states supported the development of those academies with the goal of obtaining power through knowledge: knowledge about routes, commodities, agricultural products, medicines, exotic animals, and wandering stars. The legacy of Spain to the emergence of modern science consisted in establishing a very successful model of domination based on empirical knowledge. Spanish books, institutions, and experts helped other Europeans to establish their own models of domination based on this type of knowledge, which would become the new science of the seventeenth century.

APPENDIX 1

Pilots and Cosmographers at the Casa de la Contratación

TABLE 1. CHIEF PILOTS (SIXTEENTH AND SEVENTEENTH CENTURIES)

1508[1] (March 22)	Amerigo Vespucci (d. 1512)
1512[2] (March 25)	Juan Díaz de Solís (d. 1516)
1512[3] (September 18)	Juan Vespucci
1518[4] (February 5)	Sebastian Cabot (1548 moved to England; d. 1557)
1550[5] (June 14) to 1552 (October 1)	Diego Sánchez Colchero, temporarily
1552[6]	Alonso de Chaves (d. 1587)
1586[7] (April 13)	Rodrigo Zamorano (appointed for four years; reappointed in 1590 for four years; d. 1620)
1595[8] (September 16): The crown now called a public competition to select candidates for the office of chief pilot	Pedro Ambrosio de Ondériz (already the royal cosmographer; appointed temporarily as royal pilot; d. 1596)
1596[9] (May 15)	Andrés García de Céspedes (d. 1611)
1598[10] (April 14)	Rodrigo Zamorano
1620[11] (December 29)	Diego Ramírez de Arellano (d. 1634)
1625[12] (June 22)	Antonio Moreno (d. 1634)

(*continued*)

TABLE 1. (*CONTINUED*)

1633[13] (June 22)	Francisco de Ruesta (appointed ship surveyor on December 25, 1645;[14] d. 1673)
1674[15] (June 17)	Juan Cruzado de la Cruz y Mesa (d. 1692)

TABLE 2. COSMOGRAPHERS AND MASTERS OF INSTRUMENTS

1519[16]	Nuño García de Torreño (master of instruments)
1523[17] (July 1)	Diego Ribeiro (first cosmographer of instruments; Portuguese; d. 1533)
1524[18] (May 27)	Jorge and Pedro Reinel (Portuguese; masters of instruments and charts)
1528[19] (April 4)	Alonso de Chaves (appointed chief pilot in 1552; d. 1587)
1532?[20]	Alonso de Santa Cruz (d. 1567)
1534[21] (May 21)	Diego Gutiérrez (d. 1554)
1537[22] (April 20)	Pedro Mejía (d. 1551)
1553[23] (May 18)	Sancho Gutiérrez (d. 1574)
1554[24] (October 22)	Diego Gutiérrez (replacing his father who died in this year: see above)
1579[25] (August 26)	Rodrigo Zamorano
1586[26] (October 7)	Domingo de Villarroel (in 1596 moved to France)
1593[27] (September 20)	Pedro Ambrosio de Ondériz (already chief cosmographer; special appointment to improve navigational instruments and charts; when he died in 1596, Andrés García de Céspedes continued and finished his work)
1598[28] (August 26)	Gerónimo Martín de Pradillo (d. 1602)
1603[29] (September 21)	Antonio Moreno (he competed for the office through examinations; d. 1635)
1635[30] (August 17)	Juan Herrera de Aguilar (d. 1647)
1652[31] (June 30)	Sebastián de Ruesta (brother of chief pilot Francisco de Ruesta; d. 1669)

TABLE 2. (*CONTINUED*)

1674[32] (October 6)	Miguel Suero (d. 1677)
1680[33] (December 19)	Manuel Salvador Barreto (d. 1709)

TABLE 3. COSMOGRAPHER-LECTURERS (SIXTEENTH AND SEVENTEENTH CENTURIES)

[1508][34]	Amerigo Vespucci (instruction was voluntary for pilots, not yet a requirement for their examinations; instruction was part of the duties of the chief pilot)
[1528][35]	Alonso de Chaves (instruction was still voluntary; he was assigned lectures at Hernando Columbus's house without pay)
1552[36] (December 4): Institutionalization of the instruction of pilots as an activity separate from those of the chief pilot	Jerónimo de Chaves (son of Alonso de Chaves; d. 1574)
1569[37] (May 25)	Sancho Gutiérrez (interim, replacing Jerónimo de Chaves; d. 1580)
1574[38] (May 11)	Licenciado Diego Ruiz (died without taking office)
1575[39] (November 20)	Rodrigo Zamorano (appointed for five years)
1579[40] (August 26)	Rodrigo Zamorano (five-year extension)
1584[41] (August 28)	Rodrigo Zamorano (already royal cosmographer, unlimited time-extension as lecturer of cosmography; d. 1620)
1612[42] (October 28)	Antonio Moreno (already cosmographer; d. 1635)
1635[43] (July 3): The lecturer of cosmography is divided into two positions: the lecturer of cosmography as such and the lecturer of ship management and construction	Lucas Guillén de Veas (lecturer of ship management and construction; d. 1644)

(*continued*)

TABLE 3. (*CONTINUED*)

	Juan Herrera de Aguilar (lecturer of cosmography; d. 1647)
	The Junta de Guerra de Indias decided that Lucas Guillén de Veas held both positions; Herrera was appointed cosmographer of instruments a month later[44]
1636[45] (September 16)	Rodrigo Zamorano de Oceta (son of Rodrigo Zamorano; interim, to replace Guillén de Veas during his absences; d. 1647)
1654[46] (October 15)	Francisco de Ruesta (already chief pilot and ship surveyor; interim; d. 1673)
1658[47] (December 24)	Juan de Saavedra (d. 1666)
1668[48] (March 17)	Alonso de Vacas Montoya (d. 1708)
1707[49] (September 28)	Francisco Antonio de Orbe

NOTES

1. AGI, Contratación 5784, l. 1, f. 4v.

2. Royal decree transcribed by José Pulido Rubio in *El piloto,* p. 465; see also his list of pilots and cosmographers, p. 979ff.

3. AGI, Contratación 5784, l. 1, f. 20r.

4. Ibid., ff. 26v–27v.

5. Pulido, *El piloto,* pp. 610–611.

6. Ibid., p. 609.

7. AGI, Contratación 5784, l. 3, ff. 40v–41.

8. AGI, Indiferente 426, l. 28, ff. 219v–220v. Pulido Rubio found a royal decree of June 11, 1596, appointing Ondériz as chief pilot, but by this time Ondériz had already died; see Pulido, *El piloto,* p. 277f.

9. Ibid., p. 277.

10. Ibid., p. 979.

11. Ibid.

12. Ibid.

13. AGI, Contratación 5785, l. 1, ff. 133v–135.

14. Ibid., ff. 251–252.

15. Pulido, *El piloto,* p. 979.

16. C. H. Haring, *El comercio y la navegación entre España y las Indias en época de los Habsburgos,* p. 41.

17. Pulido, *El piloto,* p. 982. In 1533 he had asked the king to be appointed for this position; see AGI, Indiferente 1204, No. 21 (2).

18. AGI, Indiferente 1204, No. 21 (3) and (4).
19. Pulido, *El piloto,* p. 982.
20. In the royal decree of November 20, 1532, Santa Cruz is already called "cosmographer"; see AGI, Indiferente 1962, l. 5, f. 41v. Pulido Rubio says that Santa Cruz was appointed on June 21, 1537; see *El piloto,* p. 982.
21. Ibid., p. 982.
22. Ibid.
23. AGI, Contratación 5784, l. 1, f. 97v.
24. AGI, Indiferente 2005.
25. AGI, Contratación 5784, l. 3, f. 3.
26. Pulido, *El piloto,* p. 982. In August 1586 he asked the king for this position; see AGI, Patronato 262, R. 1.
27. Pulido, *El piloto,* p. 270; see the instructions to Ondériz, AGI, Indiferente 742, No. 151C.
28. Pulido, *El piloto,* p. 983; AGI, Contratación 5784, l. 3, ff. 122–122v.
29. AGI, Contratación 5784, l. 3, ff. 122–122v.
30. Pulido, *El piloto,* p. 983.
31. Ibid.
32. Ibid.
33. Ibid.
34. AGI, Indiferente 1961, l. 1, ff. 65v–67.
35. AGI, Indiferente 421, l. 13, f. 295v.
36. AGI, Contratación 5784, l. 1, ff. 95–95v.
37. Ibid., ff. 158r–158v.
38. Ibid., ff. 169r–170v.
39. Ibid., ff. 170v–171.
40. Ibid., l. 3, f. 3.
41. Ibid., ff. 39r–39v.
42. Pulido, *El piloto,* p. 981.
43. AGI, Contratación 5785, l. 1, ff. 100v–103.
44. Pulido, *El piloto,* p. 101.
45. Ibid., p. 107ff.
46. Ibid., p. 101ff.
47. AGI, Contratación 5785, l. 2, ff. 105–106.
48. Pulido, *El piloto,* p. 129.
49. Ibid., p. 982.

APPENDIX 2

Instruments

TABLE 4. INSTRUMENTS TO RESCUE OBJECTS FROM THE OCEAN

Date (Y/M/D)	*Inventor, Craftsman*	*Instrument, Device, Technology*
1520/08/27	Juan de Cárdenas	Instrument for fishing oysters in Cubagua[1]
1526/06/20	Juan Fernández de Castro	Promise to make an instrument for removing gold from rivers in Cuba, Hispaniola, San Juan, and Jamaica[2]
1528/01/10	Luis de Lampiñán, son of the Conde Juan Andrea de Lampiñán	Instrument for fishing oysters[3]
1539/03/07	Nicolás de Rodas, Greek	Instrument for diving[4]
1544/09/29	Antonio Binali	Offer to make instruments and devices for rescuing a box of gold and silver lost in the ocean[5]
1565/04/01	Pedro de Herrera	Instrument for rescuing gold, silver, and oysters from the ocean[6]
1565/04/08	Juan Palomino, a clergyman	Instrument for rescuing gold and silver from the ocean[7]

TABLE 4. (*CONTINUED*)

1568/12/19	Antonio Luis de Cabrera, Antonio de Luna, and Diego de Lira	Instrument for rescuing gold and valuables from wrecked ships[8]
1573/11/02	Francisco de Soler	Diving device for retrieving objects from the ocean[9]
1577/10/21	Cristóbal Maldonado	Instrument for rescuing gold, silver, and lost objects from wrecked ships[10]
1583	José Bono (or Jusepe Vono), from Sicily	Diving device (a bell)[11]
1592/06/29	Bartolomé Francés	Instrument for fishing oysters, called a *tartana*[12]
1605/06/20	Don Jerónimo de Ayanz	Instrument for rescuing objects from the sea and fishing oysters[13]
1612/12/22	Thomas de Cardona, a Venetian resident of Seville, and Sancho de Meras	Instrument for rescuing objects from the oceans and fishing oysters[14]
1640/05/14	Diego Pérez Hidalgo	Invention for extracting gold from rivers and fishing oysters[15]

TABLE 5. MILLS

Date (Y/M/D)	*Inventor, Craftsman*	*Instrument, Device, Technology*
1569/08/24	Pedro Juan de Lastanosa, royal servant	Improved mill[16]
1577/10/21	Fulvio and Simón Genga, from Urbino	Invention of a flour mill[17]
1579/03/30	Jácome Valerio	Two mills that used human force[18]
1579/05/31	Jorge Grambono, from Rome	Mill[19]
1580/10/14	Alfonso Sánchez Cerrudo, a clergyman	Mill for metals[20]

(*continued*)

TABLE 5. (CONTINUED)

Date (Y/M/D)	Inventor, Craftsman	Instrument, Device, Technology
1587/02/18	Francisco Capuano, from Sicily, and Matía Janer, from Barcelona	Mill[21]
1607?	Pedro de Avila	Mill for metals[22]

TABLE 6. NAVIGATIONAL INSTRUMENTS

Date (Y/M/D)	Inventor, Craftsman	Instrument, Device, Technology
1528/08/21 to 1531/10/13	Diego Ribeiro, royal cosmographer	Metal bilge pumps[23]
1535/06/28	Vicente Barreros, carpenter	Offer to make wooden bilge pumps, four times cheaper and safer than those made from metal by Ribeiro[24]
1540/01/31	Martín Sánchez de Matienza	Lighter and simpler bilge pump than the current ones[25]
1559	Jácome Francisco	Device for careening cheaper and more efficiently than other devices[26]
1572/03/05	Juan Alonso	Instrument for taking altitude at any time of the day[27]
1573/02/17	Pedro Menéndez de Aviles, governor and captain general of Florida	Instrument (a compass) for finding longitudes[28]
1573/03/03	Licenciado Ruiz	Instrument for taking altitudes[29]
1573/11/10	Pablo Matía	Device for repairing ships[30]
1574/01/08	Juan de Herrera	Several navigational instruments[31]
1586/08	Don Domingo de Villarroel, clergyman and cosmographer from Naples	Navigational instrument (a sunglass)[32]
1586/12/16	Don Mateo Vázquez de Leca y Luchiano	Device for careening ships[33]

TABLE 6. (*CONTINUED*)

1592/06/10	Pedro Grateo, a watchmaker from France	Navigational instruments (asking for a license to make instruments)[34]
1599–1602	Andrés García de Céspedes	New, improved navigational instruments[35]
1609/05/09	Luis de Fonseca	Navigational instruments: a regular compass and a vertical compass[36]
1612/07/12	Juan Martines	Compass that solves the problem of the variation[37]

TABLE 7. INSTRUMENTS AND TECHNOLOGIES FOR EXPLOITING METALS

Date (Y/M/D)	*Inventor, Craftsman*	*Instrument, Device, Technology*
1543/07/06	Jordan de Meçeta	Bellows for smelting[38]
1554/05/10	Miguel Jerónimo Pallas	New method for extracting silver[39]
1554/11/17	Bartolomé de Medina	Method for extracting silver known as amalgamation[40]
1562/05/31	Pedro de Torre	Instrument for smelting gold and silver without bellows[41]
1570/05?	Benito de Morales	New method for exploiting metals[42]
1572/04/30	Francisco de Acosta	Device to make a more efficient use of mercury in the exploitation of gold and silver[43]
1578/07/19	Don Carlos de Samano, magistrate of Mexico	Invention for the exploitation of silver mines[44]
1588/08/13	Pedro de Estrella	Device for working in shafts lacking wind or light[45]
1601–1603	Antonio de Muñoz	Method for exploiting sweepings of ore[46]
1607/03/24	Don Antonio Sigler	Method for exploiting silver without mercury[47]

(*continued*)

TABLE 7. (*CONTINUED*)

Date (Y/M/D)	*Inventor, Craftsman*	*Instrument, Device, Technology*
1607/05/08	Pedro de Veraza	Device for exploiting gold[48]
1609?	Antonio de Muñoz	"Never before used invention" for exploiting silver and gold[49]

TABLE 8. OTHER INSTRUMENTS

Date (Y/M/D)	*Inventor, Craftsman*	*Instrument, Device, Technology*
1549/04/11	Baltasar de los Ríos	Instrument for illumination[50]
1572/08/07	Juan de Puy	Invention for obtaining drinking water from seawater[51]
1578/08/04	Bartolomé de Gálvez, resident of the village of Rambla	Pump for extracting water from wells, mines, and ships[52]
1588/08/13	Francisco de Torralva, from the castle of Garci Muñoz	Pump for extracting water from wells, mines, and ships[53]
1594/06/17	Don Miguel de la Cerda	Device for minting money[54]
1598/06/22	Doctor Vellerino	Device for minting money in Quito (related to Don Miguel de la Cerda's device)[55]
1601/05/22		Mill for niters (gunpowder)[56]

NOTES

1. AGI, Indiferente 420, L. 8, ff. 253v–255r.
2. AGI, Indiferente 421, L. 11, ff. 41r–42r; Indiferente 421, L. 12, ff. 47r–47v.
3. AGI, Panamá, 234, L. 3, f. 56v; *CODOIN*, series 2, vol. 24, p. 152; Justicia, 7, No. 4.
4. AGI, Indiferente 1962, L. 6, ff. 176r–176v.
5. AGI, Indiferente 1093, R. 3, No. 56.
6. AGI, Indiferente 425, L. 24, ff. 228r–229r; and Indiferente 427, L. 30, ff. 190r–190v.
7. AGI, Indiferente 425, L. 24, ff. 230v–232v.
8. Ibid., ff. 423v–424r.
9. AGI, Indiferente 426, L. 25, ff. 256r–257r.

10. Ibid., L. 26, ff. 59r–60v.; Indiferente 426, L. 26, f. 109v.
11. AGI, Patronato 260, No. 1, R. 10.
12. AGI, Escribanía 1008c.
13. AGI, Patronato 171, No. 1, R. 37.
14. AGI, Indiferente 428, L. 34, ff. 62r–68r.
15. AGI, Indiferente 429, L. 38, ff. 160–161.
16. AGI, Indiferente 426, L. 25, ff. 17r–18v.
17. Ibid., L. 26, ff. 58r–59r.
18. Ibid., ff. 159r–159v.
19. Ibid., ff. 165v–166v.
20. Ibid., ff. 218v–220r.
21. Ibid., L. 27, ff. 154v–156r.
22. AGI, Charcas 84, No. 2.
23. In 1528 Hernando Columbus wrote to the crown about Ribeiro's pumps; see AGI, Indiferente 421, L. 13, f. 295r/v.
24. AGI, Indiferente 1961, L. 3, ff. 293v–294; and Indiferente 1204, No. 53.
25. AGI, Indiferente 1206, No. 30.
26. AGI, Indiferente 738, No. 127.
27. AGI, Patronato 264, R. 3.
28. AGI, Indiferente 426, L. 25, ff. 226r–227v.
29. Ibid., ff. 228v–229v.
30. Ibid., ff. 259v–262r.
31. AGI, Patronato 259, R. 58.
32. AGI, Patronato 262, R. 1.
33. AGI, Indiferente 426, L. 27, ff. 146r–146v.
34. AGI, Contratación 734, No. 1.
35. AGI, Patronato 262, R. 2.
36. AGI, Patronato 262, R. 4.
37. Ibid., R. 5.
38. AGI, Indiferente 423, L. 20, ff. 645v–646v.
39. AGI, Indiferente 424, L. 21, ff. 333v–334v.
40. Archivo General de la Nación, México, Ramo de Mercedes, V, ff. 87–87v, transcribed in Luis Muro, "Bartolomé de Medina," p. 518ff.
41. AGI, Indiferente 425, L. 24, ff. 105r–105v.
42. AGI, Indiferente 856. This document does not have a date, but there is a reference to Benito de Morales and his new method in Consulta del Consejo, May 24, 1570, Madrid, AGI, Indiferente 738, No. 119.
43. AGI, Indiferente 426, L. 25, ff. 178r–179r.
44. Ibid., L. 26, ff. 111v–112.
45. Ibid., L. 28, ff. 5–6.
46. AGI, Indiferente 1800.
47. AGI, Indiferente 428, L. 33, ff. 9v–10v.
48. AGI, Quito 27, No. 7.
49. AGI, Indiferente 856.
50. AGI, Indiferente 424, L. 21, ff. 333v–334v.
51. AGI, Indiferente 426, L. 25, ff. 193v–194r.

52. Ibid., L. 26, ff. 115v–116v.
53. Ibid., L. 28, ff. 4r–5r.
54. AGI, Indiferente 742, No. 173.
55. AGI, Quito 1, No. 75.
56. AGI, México 24, No. 58.

APPENDIX 3

Spanish Scientific Books

Author	Book	Edition
Martín Fernández de Enciso	*Suma de geografía que trata de todas las partidas y provincias del mundo: en especial de las indias y trata largamente del arte del marear juntamente con la espera en romance: con el regimiento del sol y del norte: agora nuevamente emendada de lgunas defectos que tenia la impression passada*	Seville, 1519 and 1530 partial English trans., 1578
Felipe Guillén de Castro	"Carta al Rey de Portugal," about longitude, declination, compass	Seville, 1525
Francisco Falero	*Tratado del Esphera y del arte de marear con el regimiento de las alturas, con algunas reglas nuevamente escritas muy necesarias*	Seville, 1535
Pedro de Medina	*Arte de navegar en que se contienen todas las reglas, declaraciones, secretos y avisos, que a la buena navegación son necessarios, y se deven saber*	Valladolid, 1545

(*continued*)

Author	Book	Edition
Jerónimo de Chaves	*Chronographia o Repertorio de los tiempos*	Seville, 1548
Martín Cortés	*Breve compendio de la sphera y de la arte de navegar, con nuevos instrumentos y reglas, exemplificado con muy subtiles demostraciones*	Seville, 1551
Pedro de Medina	*Regimiento de navegación. En que se contienen todas las reglas, declaraciones y avisos del libro del arte de navegar*	Seville, 1552
Bernardo Pérez de Vargas	*Repertorio perpetuo*	Toledo, 1563
Juan Pérez de Moya	*Arte de navegar*	Manuscript, 1564
Alonso de Santa Cruz	*Islario general* and the *Libro de las longitudes*	Manuscript, 1566
Juan Pérez de Moya	*Tratado de Mathematicas en que se contienen cosas de Arithmetica, Geometria, Cosmographia, y Philosophia natural. Con otras varias materias, necesarias a todas las artes Liberales y Mechanicas*	Alcalá de Henares, 1573
Nicolás Bautista Monardes	*[Primera y Segunda y Tercera partes de la] Historia Medicinal de las cosas que se traen de nuestras Indias Occidentales que sirven en Medicina*	Seville, 1574
Rodrigo Zamorano	*Compendio del arte de navegar*	Seville, 1581
Francisco Vicente de Tornamira	*Chronologia y repertorio de los tiempos*	Pamplona, 1585
Rodrigo Zamorano	*Cronologia y repertorio de la razón de los tiempos*	Seville, 1585
Andrés de Poza	*Hydrografia la mas curiosa que hasta aqui ha salido a la luz, en que de mas de un derrotero*	Bilbao, 1585

	general, se enseña, la navegación por altura y derrota, y la del Este Oeste: con la Graduación de los puertos, y la navegación al Catayo por cinco vias diferentes	
Diego García de Palacio	*Instruccion nauthica para el buen uso y regimiento de las Naos, su traça, y govierno conforme a la altura de Mexico*	Mexico City, 1587
José de Acosta	*Historia natural y moral de las Indias*	Seville, 1590 Barcelona, 1591 Madrid, 1608 Madrid, 1614 Trans.: French (1598, 1600, 1606, 1616, 1617); Italian (1596); Dutch (1598, 1624); German (selections: 1598, 1600, 1605); English (1604)
Simón Tovar	*Examen i censura del modo de averiguar las Alturas de las tierras, por la altura de la Estrella del Norte, tomada con la Ballestilla. En que se demuestran los muchos errores que ay en todas las Reglas, que para esto se an usado hasta agora: y se enseñan las que conviene usarse; y guardarse en nuestros tiempos; y el modo como podra hazerse en los venideros*	Seville, 1595
Pedro de Siria	*Arte de la verdadera navegación*	Valencia, 1602
Andrés García de Céspedes	*Regimiento de navegación*	Madrid, 1606
Francisco Suárez de Argüello	*Ephemerides generales de los movimientos de los cielos por doze años, desde el de MDCVII*	Madrid, 1608 [written in 1582]

(*continued*)

Author	Book	Edition
	hasta el de MDCXVIII segun el Serenisimo Rey don Alonso en los quatro Planetas inferiores y Nicolao Copernico en los tres superiores que mas conforma con la verdad y observaciones, como se dira en el prologo	
Andrés Río y Riaño	*Tratado de un instrumento por el cual se conocerá la nordestacion, o noruestacion de la Aguja de marear, navegando por la mayor altura del sol, o de otra estrella; o por dos alturas iguales: y de la utilidad que se ha de seguir*	Seville, 1608
Antonio de Herrera	*Historia de los hechos de los Castellanos en las islas y Tierra-firme del mar Oceano con una descripcion de las Indias orientales y sus mapas*	Madrid, 1601–1615
Lorenzo Ferrer Maldonado	*Imagen del mundo, sobre la esfera, cosmografia, y geografia, teorica de planetas, y arte de navegar*	Alcalá de Henares, 1626

NOTE

Note: For a complete list of books, see José Simón Díaz, *Bibliografía de la literatura hispánica;* Antonio Palau y Dulcet, *Manual del librero hispano-americano;* Felipe Picatoste y Rodríguez, *Apuntes para una biblioteca científica española del siglo XVI, estudios biográficos y bibliográficos de ciencias exactas, físicas y naturales y sus inmediatas aplicaciones en dicho siglo;* and José Luis Valverde, Teresa Bautista, and María Teresa Montaña, *Libros de interés histórico-médico-farmacéutico conservados en la Biblioteca de la Real Academia de Medicina de Sevilla.*

Notes

INTRODUCTION

1. Consulta del Consejo, October 9, 1592, Madrid, Archivo General de Indias (from now on AGI), Indiferente, 746, No. 44.

2. For more on this theme, see Antonello Gerbi, *Nature in the New World;* Raquel Alvarez Peláez, *La conquista de la naturaleza americana;* Anthony Pagden, *The Fall of Natural Man;* and John H. Elliott, *The Old World and the New.*

3. My approach to Spanish explorers' encounters with the New World regarding empirical practices is similar to the one taken by Stephen Greenblatt regarding the conceptualization of wonder and the New World. For him, "the frequency and intensity of the appeal to wonder in the wake of the great geographical discoveries of the late fifteenth and early sixteenth centuries helped (along with many other factors) to provoke . . . [the] conceptualization [of the marvelous]." In my case, the Spanish encounter with the New World intensified and expanded institutions and practices regarding personal experience and empirical information. See Stephen Greenblatt, *Marvelous Possessions,* p. 19.

4. Carolyn Merchant, *The Death of Nature: Women, Ecology and the Scientific Revolution;* Urs Bitterli, *Cultures in Conflict;* William J. Bouwsma, "Anxiety and the Formation of Early Modern Culture," in *After the Reformation: Essays in Honor of J. H. Hexter,* ed. Barbara C. Malament; Anthony Grafton, *New Worlds, Ancient Texts: The Power of Tradition and the Shock of Discovery;* Elliott, *The Old World and the New.*

5. Paula Findlen, *Possessing Nature: Museums, Collecting, and Scientific Culture in Early Modern Italy,* p. 58f.

6. See Allen G. Debus, *The Chemical Philosophy: Paracelsian Science and Medicine in the Sixteenth and Seventeenth Centuries,* vol. 1, p. 61ff.

7. See Pamela H. Smith, *The Business of Alchemy* (Princeton, 1994); Tara E. Nummedal, "Practical Alchemy and Commercial Exchange in the Holy Roman Empire," in *Merchants and Marvels: Commerce, Science, and Art in Early Modern Europe,* ed. Pamela H. Smith and Paula Findlen, p. 207ff.

8. The following account is taken from several sources on the history of Spain. See Antonio Domínguez Ortiz, *The Golden Age of Spain, 1516–1659;* Antonio Domínguez Ortiz, "Instituciones políticas y grupos sociales en Castilla durante el siglo XVII," in *Instituciones y sociedad en la España de los Austrias;* John H. Elliott, "The Decline of Spain," *Past and Present* 20 (1961); John H. Elliott, *Imperial Spain, 1469–1716;* Manuel Fernández Alvarez, *La sociedad española del Renacimiento;* Manuel Fernández Alvarez, *La sociedad española en el Siglo de Oro;* Clarence Henry Haring, *Trade and Navigation between Spain and the Indies;* Henry Kamen, *Philip of Spain;* John Lynch, *The Hispanic World in Crisis and Change, 1598–1799;* John Lynch, *Spain, 1516–1598: From Nation State to World Empire;* José Antonio Maravall, *Estado moderno y mentalidad social (siglos XV a XVII),* vol. 1; and I. A. A. Thompson, *War and Society in Habsburg Spain: Select Essays.*

9. José Losana Méndez, *La sanidad en la época del descubrimiento de América,* p. 35.

10. See, for instance, two surveys by leading scholars in the field that barely (if at all) mention Spain: Peter Dear, *Revolutionizing the Sciences: European Knowledge and Its Ambitions, 1500–1700;* and Steven Shapin, *The Scientific Revolution.*

11. For more on Medina and Cortés, see Francisco José González, *Astronomía y navegación en España: Siglos XVI–XVIII,* pp. 72 and 78; on Borough, see Peter Barber, "England II: Monarchs, Ministers, and Maps, 1550–1625," in *Monarchs, Ministers, and Maps: The Emergence of Cartography as a Tool of Government in Early Modern Europe,* ed. David Buisseret; José López Piñero and José Pardo Tomás, "The Contribution of Hernández to European Botany and Materia Medica," in *Searching for the Secrets of Nature: The Life and Works of Dr. Francisco Hernández,* ed. Simon Varey, Rafael Chabrán, and Doris B. Weiner.

12. For more on the increased materialism of sixteenth- and seventeenth-century Europe and the development of modern scientific practices, see Chandra Mukerji, *From Graven Images: Patterns of Modern Materialism,* Chapters 3 and 4.

13. Alvarez Peláez, *La conquista;* David C. Goodman, *Power and Penury: Government, Technology, and Science in Philip II's Spain;* José María López Piñero, *Ciencia y técnica en la sociedad española de los siglos XVI y XVII;* and José María López Piñero, ed., *Historia de la ciencia y de la técnica en la corona de Castilla,* vol. 3.

14. Robert K. Merton, "STS: Foreshadowing of an Evolving Research Program in the Sociology of Science," in *Puritanism and the Rise of Modern Science: The Merton Thesis,* ed. Bernard Cohen, p. 361.

15. Edgar Zilsel, "The Sociological Roots of Science," in *The Social Origins of Modern Science,* ed. Diederick Raven et al.

16. For a discussion on the Casa de la Contratación and its Chamber of Knowledge, see Chapter 2.

17. For more on the different contexts for the study of institutionalized science and preinstitutionalized science, see Steven Shapin, who makes a similar distinction but in terms of professionalized and preprofessionalized science: "History of Science and Its Sociological Reconstructions," *History of Science* 20 (1982).

18. Lucile H. Brockway, *Science and Colonial Expansion: The Role of the British Royal Botanic Gardens;* Mary Louise Pratt, *Imperial Eyes: Travel Writing and Transculturation.*

19. Quoted in Melquiades Andrés Martín et al., eds., *El siglo del Quijote (1580–1680)* (Madrid: Espasa-Calpe, 1996).

20. Agustín Farfán, *Tractado breve de medicina.* The first edition is from 1579. The 1592 edition is the second edition.

21. Antonio de León Pinelo and Andrés González de Barcía, *Epitome de la biblioteca oriental y occidental, náutica y geográfica: Añadido y enmendado nuevamente . . . ,* vol. 2, p. 899.

22. I am already working on a manuscript entitled "Translating Books, Observing Nature: Spanish Science and English Reports from the New World" that examines these connections.

1. SEARCHING THE LAND FOR COMMODITIES

1. Real Cédula de la Reina a las justicias de sus reinos, April 5, 1530. Madrid, AGI, Indiferente 422, L. 14, f. 67v. The queen was Isabel de Portugal, wife of Charles V, who assumed the regency while Charles traveled to Italy for the imperial coronation. All translations are mine unless otherwise mentioned.

2. See a version of the Santo Domingo balsam case in Antonio Barrera, "Local Herbs, Global Medicines: Commerce, Knowledge, and Commodities in Spanish America," in *Merchants and Marvels,* ed. Smith and Findlen.

3. On this early modern global setting or, as Richard H. Grove calls it, "global framework of trade and travel," see Richard H. Grove, *Green Imperialism,* p. 3ff.

4. Edmund Burke, "Contested Paradigms in Early Modern World History," paper presented at the University of California at Davis conference on Modernity's Histories in Global Context: Contested Narratives, Models, Processes, May 17–18, 1997.

5. See Kenneth Pomeranz, *The Great Divide: China, Europe, and the Making of the Modern World Economy,* p. 44ff.

6. For more on the House of Trade, see Chapter 2 below; on the Council of Indies, see Ernst Schäfer, *El Consejo Real y Supremo de las Indias;* and his "El origen del Consejo de Indias," *Investigación y Progreso* 7, no. 5 (May 1933): 141–145.

7. Andrés Laguna, *Pedacio Dioscórides Anazarbeo, acerca de la materia medicinal, y de los venenos mortíferos: Traduzido de la lengua Griega en la vulgar Castellana, e illustrado con claras y substanciales annotaciones, y con las figuras de unnúmeras plantas exquisitas y raras por el doctor . . . , médico de Julio III, Pontífece Máximo,* f. 26ff. On the importance of balsam for sixteenth-century naturalists, see Findlen, *Possessing Nature,* p. 270ff.

8. Pero Tafur, *Andanças e viajes de Pero Tafur por diversas partes del mundo avidos (1435–1439),* pp. 85–86.

9. Ibid., p. 575, note: *bálsamo.*

10. Ramón Carande, *Carlos V y sus banqueros,* vol. 2, p. 13ff.

11. Real Cédula del Rey Don Carlos, January 29, 1525, Madrid, AGI, Contratación 5787, No. 1, L. 1. ff. 33–34v.

12. Oviedo commented that this was not the real balsam but something different that Villasante called balsam. Oviedo also stated that Villasante learned the secret of balsam either from his *cacica* wife or from an Italian physician who went to the Indies in 1515 and died there; see Gonzalo Fernández de Oviedo y Valdés, *Historia general y natural de las Indias,* vol. 2, p. 11.

13. Ernst Schäfer, "Antonio de Villasante, descubridor droguista en la isla Española," *Investigación y Progreso* 9, no. 1 (1935): 13. Villasante's name appears in a document signed in Santo Domingo in February 1515. In this document, Villasante was proposed as a witness (together with other residents) to answer questions about Rodrigo de Albuquerque's activities on the island in 1514. AGI, Justicia 1003, transcribed in Luis Arranz Márquez, *Repartimiento y encomiendas en la Isla Española (El Repartimiento de Albuquerque de 1514).*

14. Arranz Márquez, *Repartimiento,* p. 560. On the Tainos, see Irving Rouse, *The Tainos: Rise and Decline of the People Who Greeted Columbus.* Cook suggests that the Taino population around 1492 might have been half a million; by 1518–1519 it had fallen to around 18,000; and by 1542 the native population was less than 2,000. Villasante's information came from a group that was disappearing from the earth. See David N. Cook, *Born to Die: Disease and New World Conquest, 1492–1650,* pp. 23–24.

15. Real Provisión proponiendo un asiento con Antonio de Villasante sujeta a la presentación de un reporte de Villasante sobre el bálsamo y otras drogas, April 4, 1528, Madrid, AGI, Indiferente 421, L. 13, ff. 85r–86v.

16. Relación de Antonio de Villasante, n.d. (but it was probably presented in mid-1528), AGI, Indiferente 857. On April 4, 1528, the king ordered Villasante to present a report before the council. By June 14, 1528, he had already submitted his report; see Indiferente 421, L. 13, ff. 213v–214r. Ernst Schäfer thinks that this document dates from around 1526; according to him, Villasante was in Spain in 1525; see "Antonio de Villasante," p. 14. Perhaps Villasante was in Spain in 1525 or 1526 and at that time sought support for his project. The call number given by Schäfer for Villasante's report, Indiferente 856, is a mistake; it is Indiferente 857; see also a document in *Colección de documentos inéditos relativos al descubrimiento, conquista y colonización de las posesiones españolas en América y Oceanía* (*CODOIN*), series 2, vol. 14, p. 31.

17. Villasante, Indiferente 857.

18. Ibid.

19. Instrucciones a Dr. Francisco Hernández, January 1, 1570, Madrid, AGI, Lima 569, L. 13, ff. 97v–101r. See an English translation in Simon Varey, ed., *The Mexican Treasury: The Writings of Dr. Francisco Hernández,* p. 46.

20. Nicolás Bautista Monardes, *[Primera y segunda y tercera partes de la] Historia medicinal de las cosas que se traen de nuestras Indias Occidentales que sirven en medicina,* f. 9r.

21. See María Justina Sarabia Viejo, *Don Luis de Velasco, virrey de Nueva España (1550–1564),* pp. 403–405. The Council of Indies, however, considered this project unfeasible; see Consulta del Consejo, March 21, 1559, Valladolid, AGI, Indiferente 738, No. 47.

22. Villasante, Indiferente 857: "La maña que hasta agora yo he tenydo en el sacar del licor con otros cosas de estos arvoles asy lo? que con un cuchillo cortados los rramos destos arboles con su hoja y grano y con la mano arrancaba los granos y tambien la hojarada cosa por sy y tomaba los rramos asy mondos y tambien tomava de la corteza de lo grueso del arvol hacia el tronco y lo desmenuzava y . . . taba? y lo majava encima de unas piedras o losas con otras piedras o madero despues de picado con cuchillo y asy majado lo ponya en unas vasijas de barro de? m . . . ? de barreno?nes? o labrillos? y . . . ? calentaba en un caldero con una cantidad de agua competente y la echava en el dicho barreno? y desde a un poce despues de enpapado y enbevido en el agua lo apretavba en un tornyllo de madera y sacaba dello todo el çumo y . . . d? que tenyz y lo colava y colado lo ponya en un caldero pequeno y despues tomaba otro caldero grande lleno de ceniza hasta la mytad del . . . ? / y dentro de aquel caldero de ceniza ponya y asentaba el otro caldero pequeño con el dicho licor del balsamo colado y ponya fuego debajo del caldero de la ceniza de maña que el calor dela ceniza consumyese el agua que estaba en dicho licor hasta tanto que se espesava y tornaba del color y maña que yo ho he tenydo y entregado a su mag/. . . ."

23. Ibid.

24. Ibid.: Villasante explained that balsam was "en la verdad provechoso asi en las Indias donde lo experimente muchas veces como algunas en estos reinos en sevilla y en la corte y pues para estas enfermedades que he dicho ha parecido por experiencia ser provechoso adelante podra parecer por experiencias o por relacion de los medicos si aprovechara a otras cosas y tambien ellos diran la forma que se podra tener para mas perfeccion del dicho licor y balsamo y otras cosas del dicho arbol."

25. Garciperez Morales, *Tratado del Balsamo y de sus utilidades para las enfermedades del cuerpo humano: Compuesto por el Doctor . . . cathedratico de prima en el colegio de Sancta Maria de Jesu de la ciudad de Sevilla. Dirigido al yllustrissimo señor don Pedro Giron Duque y Conde de Ureña,* f. 2r.

26. See Juan Jiménez Castellanos y Calvo Rubio, "Prólogo," in Monardes, *Historia medicinal,* pp. v–xi. For a discussion of Monardes's work, see José María López Piñero, "Las 'nuevas medicinas' americanas en la obra (1565–1574) de Nicolás Monardes," *Asclepio* 42, no. 1 (1990): 3–67. The work of Monardes was translated into English by John Frampton as Nicolás Monardes, *Ioyfull Newes Out of the Newe Founde Worlde* (London, 1577) as well as into French, Latin, and Italian.

27. Real Provisión a Antonio de Villasante, April 22, 1528, AGI, Indiferente 421, L. 13, ff. 110r–111r and 111r–112r; Real Provisión a Antonio de Villasante, June 14, 1528, AGI, Indiferente 421, L. 13, ff. 213v–214r.

28. Real Cédula a los oficiales de la Casa de la Contratación, December 16, 1513, Madrid, AGI, Panamá 233, L. 1, f. 126r.

29. Real Cédula al licenciado Rodrigo Figueroa, juez de residencia de la isla Española, July 26, 1519, Barcelona, AGI, Indiferente 420, L. 8, f. 97v.

30. Real Cédula a los oficiales de la Española, September 14, 1526, Granada, AGI, Indiferente 421, L. 11, ff. 202v–203r.

31. Carta del licenciado Barreda al rey Carlos V, October 26, 1528, Santo Domingo de la Española, AGI, Patronato 174, R. 43.

32. Carta de Barreda, AGI, Patronato 174: "la virtud mas principal que se halla en el dicho licor / es restreñir la sangre en las llagas frescas sobre ellas aplicado / y dado por la boca el fluxo de sangre por abaxo / desta virtud agora se llame opilativa que sua viscositate aut g?oficie inplendo venari orificia rectineat sanguyneus / agora sea constrictiva que sua frigiditate / r? stiticitate? constringat venas. digo que entanta manera aprieta que puesto sin ligadura parece el miembro estar atado // pues donde se vido ni en que libros se hallo tener el balssamo esta virtud antes de todo en todo contraria en lo qual por ser muy manifiesto dexo de ser prolixo//."

33. Laguna, *Dioscórides*, pp. 26–27.

34. See ibid.; Monardes, *Historia medicinal*, f. 9ff.; Conrad Gesner, *Evonymus C. Gesneri Medici de Remedis secretis, Liber Physicus, Medicus & partim etiam Chymicus, & Oeconomicus in vinorum diversi saporis apparatu, Medicis & Pharmacopolis omnibus praecipue necessarius, nunc primum in lucem editus* (this seems to be the edition from Zurich, ca. 1565; for the date and place, see Klaus Wagner, *Catálogo abreviado de las obras impresas del siglo XVI de la Biblioteca Universitaria de Sevilla*), ff. 131r–v; Pedrarias de Benavides, *Secretos de Chirurgia, especial de las enfermedades de Morbo galico y Lamparones y Mirrarchia, y asimismo la manera como se curan los Indios de llagas y heridas y otras passiones en las Indias, muy util y provechoso para en España y otros muchos secretos de chirurgia hasta agora no escritos*, ff. 30v–31r.

35. Carta de Barreda, AGI, Patronato 174, R. 43.

36. Ibid.

37. Real Cédula de la Reina a las justicias de sus reinos, April 5, 1530, Madrid, AGI, Indiferente 422, L. 14, f. 67v.

38. Ibid.

39. Ibid., f. 68r.

40. Real Cédula a los visitadores del Hospital del Cardinal de la ciudad de Toledo, April 5, 1530, Madrid, AGI, Indiferente 422, L. 14, f. 72v.

41. Real Cédula a los visitadores de varios hospitales, April 5, 1530, Madrid, AGI, Indiferente 422, L. 14, f. 72v.

42. Real Cédula al bachiller Andrés de Jodar médico, vecino de Baeza, April 5, 1530, Madrid, AGI, Indiferente 422, L. 14, ff. 73r–74v.

43. Real Cédula a varios médicos y cirujanos, April 5, 1530, Madrid, AGI, Indiferente 422, L. 14, ff. 73r–74v.

44. Real Cédula a Pedro Benito de Basniana y Franco Leardo para que puedan subir los salarios asignados a los médicos que contribuyen a la propaganda del bálsamo, July 12, 1530, Madrid, AGI, Indiferente 422, L. 14, ff. 102r–103r.

45. Real Cédula a los oficiales de Cuéllar, October 16, 1532, Madrid, AGI, Indiferente 422, L. 15, ff. 197v–198r.

46. Ibid.; Real Cédula a Diego de la Haya para que pague a Melchor de Angulo, November 27, 1532, Madrid, AGI, Indiferente 422, L. 15, f. 199v.

47. Real Cédula a Juan de Vargas para que venga a la corte, November 21, 1532, Madrid, AGI, Indiferente 422, L. 15, f. 199r.; and Mandamiento a Diego de la Haya para que pague a Juan de Vargas por haber estado en la corte, February 27, 1533, Madrid, AGI, Indiferente 422, L. 15, f. 199r.

48. Mandamiento a Diego de la Haya para que pague a Juan de Vargas por haber estado en la corte, February 27, 1533, Madrid, AGI, Indiferente 422, L. 15, f. 199r.; Real Cédula a Diego de la Haya para que pague cierta suma a Juan de Vargas, October 3, 1533, Monzón, AGI, Indiferente 422, L. 16, f. 43v; Real Cédula a Juan de Vargas, April 18, 1534, Toledo, AGI, Indiferente 422, L. 16, f. 75v.

49. Real Cédula a los alcaldes ordinarios de la villa de Amusco, May 23, 1539, Toledo, AGI, Indiferente 423, L. 19, ff. 247–248.

50. For more on this topic, see Grove, *Green Imperialism,* p. 32ff.

51. Cecil Jane, *The Four Voyages of Columbus,* vol. 1, p. 83: "Y esta provisyón ha de durar hasta que acá [en las Indias] se aya fecho cimiento de lo que acá [en España] se sembrare y plantare, digo de trigos y cevadas y viñas."

52. There were agricultural experiments before the encounter with the New World, but it was there that these experiments created quantitative problems for merchants, royal officials, and inhabitants of the New World in general. Even today people experiment with the adaptation of wheat to new environments—only now those new environments are outside earth. Scientists recently tried to grow wheat in the space-station *Mir;* the wheat they produced was infertile. See Shannon W. Lucid, "Six Months on Mir," *Scientific American* 278, no. 5 (May 1998): 46–56.

53. Letter from Zauzo to Charles V in Marcos Jiménez de la Espada, *Relaciones geográficas de Indias: Perú,* vol. 1, p. 11f.

54. Ibid., p. 12.

55. Real Cédula a los oficiales de la Isla Española, May 7, 1519, Barcelona, AGI, Panamá 233, L. 1, f. 230r.

56. Bernabé Cobo, *Obras,* vol. 1, p. 375ff.

57. In her case study of the Valle de Mexquital, Mexico, Elinor G. K. Melville explains: "Whenever ungulates (herbivores with hard horny hooves) are faced with more food than is needed to replace their numbers in the next generation," as was the case with European animals in the New World, "an ungulate irruption is the result." European animals, at least, found a paradise of food in the New World. With the exception of Peru, they did not face any strong competition from other animals.

The new animals at first reduced the standing crop of vegetation to its lowest level of density, height, and diversity. The supply of vegetation could not sustain the huge new animal population, and the population thus decreased again until the plant communities had room to recover. Animal and plant communities then reached a level of more stable accommodation. Elinor G. K. Melville, *A Plague of Sheep,* p. 6ff.

58. Real Provisión a los residentes de San Juan del rey don Ferdinand, September 27, 1514, AGI, Indiferente 419, L. 5, ff. 248v–249v.

59. Justo Lucas de Río Moreno, *Los inicios de la agricultura europea en el Nuevo Mundo (1492–1542),* pp. 49–69.

60. Ibid.; Melville, *Plague of Sheep;* Alfred Crosby, *The Columbian Exchange: Biological and Cultural Consequences of 1492;* Alfred Crosby, *Ecological Imperialism: The Biological Expansion of Europe, 900–1900.*

61. Real Cédula al gobernador de Tierra Firme, February 20, 1534, Toledo, in *Colección de documentos y manuscriptos compilados por Fernández de Navarrete,* vol. 13, part 2, p. 1101 (hereafter cited as *Colección Navarrete*).

62. Carta de Pascual de Andagoya, October 22, 1534, Tierra Firme, in *Colección Navarrete,* vol. 13, part 2, p. 1106.

63. For the names of these two Germans, see Jean-Pierre Berthe, "El cultivo del 'pastel' en Nueva España," *Historia Mexicana* 9, no. 3 (1960): 346. He suggests that Alberto Cuon could be Albrecht Cohen or Kuhn.

64. Asiento otorgado a Enrique Ynguer and Alberto Cuon, March 1527, Belpuche, AGI, Contaduría 672, No. 5; also see Berthe's analysis in "El cultivo," pp. 343–347.

65. Ibid. The Spanish name *tuza* comes from Nahuatl *totzan;* see Francisco Santamaría, *Diccionario de mejicanismos,* p. 1097.

66. Charles had given authorization to the Genovese and Germans to travel to the New World but secretly sent instruction to the Casa officials to obstruct the movement of foreigners to the New World. See Carande, *Carlos V,* vol. 1, p. 266.

67. Real Cédula a los oficiales de la Casa de la Contratación, December 30, 1537, Valladolid, AGI, Indiferente 1962, L. 5, ff. 309v–310r. I found a license to pass to New Spain granted to three of them (their names are in Spanish): Juan Barta, Domingo de San Pablo, and Bartolo de Rigazo, "natives of Toulouse and masters of making pastel"; see AGI, Pasajeros L. 2, E. 3784.

68. There were also some Portuguese administrators in New Spain. See Berthe, "El cultivo," p. 365, note 29. On Juan Bartholo (Juan Barta), whose nationality and profession Berthe does not know, see "El cultivo," note 5.

69. Berthe, "El cultivo," p. 355, cuadro 4. Juan Ximénez and Gonzalo Gómez brought some pastel "para ensayes."

70. Nineteen indigenous towns with their jurisdictions were involved in the cultivation and manufacturing of pastel in Mexico. See the Relaciones de los pueblos del pastel from April 1545 in AGI, Contaduría 672, No. 5; see also Berthe's analysis of these documents in "El cultivo," p. 349f.

71. See Merced otorgada por don Luis de Velasco a Francisco Mirantes por cuatro años para el uso de un ingenio de moler metales, December 15, 1550, Mexico, Library

of Congress, Manuscripts, Kraus Collection, item 140, ff. 10v–11r; Merced otorgada por don Luis de Velasco a Castañon de Aguero por cuatro años para el beneficio de metales, February 29, 1551, Mexico, Library of Congress, Manuscripts, Kraus Collection, item 140, ff. 33r–v. On this topic, see Chapter 3.

72. Real Cédula a Marco de Ayala, November 1, 1562, Madrid, AGI, México 2999, L. 2, ff. 6r–7r. See also the Relación de méritos y servicios de Marco de Ayala, September 15, 1561, Mérida de Yucatán, AGI, Patronato 64, R. 7.

73. Carta de Pedro de Ledesma al Rey, May 22, 1563, Mexico City, AGI, México 168. By the end of the sixteenth century indigo was coming not only from New Spain but from Honduras, the Caribbean, and Tierra Firme; see Huguette Chaunu and Pierre Chaunu, *Séville et l'Atlantique, (1504–1650)* vol. 6/2, pp. 988–993. By the early seventeenth century indigo was already exported to the Netherlands; see N. W. Posthumus, *Inquiry into the History of Prices in Holland,* vol. 1, pp. 415–418.

74. Carta de Juan Rodríguez de Noriega sobre el tinte del palo de Yucatán, April 28, 1564, n.p., AGI, Indiferente 1093, R. 11, No. 239.

75. Real Cédula al gobernador de la provincia de Yucatán, June 25, 1565, El Escorial, AGI, México 2999, L. 2, ff. 34r–34v.

76. François Chevalier, *Land and Society in Colonial Mexico,* pp. 73–74.

77. Carta real al virrey de Nueva España, don Martín Enriques, December 3, 1576, Madrid, AGI, México 109; Carta real al gobernador de Yucatán, December 3, 1576, Madrid, AGI, México 2999, L. 3, ff. 29v–31v.

2. A CHAMBER OF KNOWLEDGE

1. Richard Hakluyt, *The Principal Navigations, Voyages, Traffiques & Discoveries of the English Nation,* vol. 1, p. xxxv.

2. See the Real Provisión a Americo Vespuccio, August 6, 1508, Valladolid, AGI, Indiferente 1961, L. 1, ff. 65v–67.

3. Haring, *Trade and Navigation,* p. 35.

4. From the translations of Richard Eden in the 1550s to the translations of Samuel Purchas in the 1620s, the English had access to Spanish books on natural history, navigation, and medicine by Peter Martyr d'Anghiera, Gonzalo Fernández de Oviedo, Martín Cortés, Francisco López de Gómara, and Nicolás Monardes. As mentioned in the introduction, this is the subject of my current book project.

5. Fernand Braudel, *The Mediterranean and the Mediterranean World in the Age of Philip II;* Manuel Fernández Alvarez, *Felipe II y su tiempo.*

6. For more on las Casas's arguments and the Valladolid debates, see David Brading, *Orbe Indiano: De la Monarquía católica a la República criolla, 1492–1867;* Lewis Hanke, *The Spanish Struggle for Justice in the Conquest of America.*

7. Richard E. Greenleaf, *Inquisición y sociedad en el México colonial.*

8. For an account of coastal navigation practices, see Braudel, *The Mediterranean,* vol. 1, p. 103ff.; see also J. H. Parry, *The Discovery of the Sea,* Chapter 2.

9. Martín Fernández de Enciso, *Suma de geografía que trata de todas las partidas y provincias del mundo: en especial de las indias y trata largamente del arte del marear juntamente con la espera en romance: con el regimiento del sol y del norte: agora nuevamente emendada de algunos defectos que tenia la impression passada,* f. 22r.

10. Iberian Arabs developed a strong scholarly tradition around Ptolemy's *Almagest.* In the twelfth century al-Bitruji (Alpetragio) transformed Ptolemy's system of eccentrics and epicycles by reverting to the earlier theories of spheres endorsed by Aristotle. The astronomical treatise of al-Farghani (Alfragano or Alfraganus) relied very much on Ptolemy's treatise; and his work became one of the main sources of Ptolemaic astronomy until the Renaissance. Al-Battani (Albategnio or Albategnius) wrote a treatise that modified Ptolemy's system to incorporate new astrological observations. The scholar Jabir ibn Aflah also revised the work of Ptolemy; he exposed the inconsistencies of Ptolemy's system with new and more accurate observations in his treatise *Islah al-magisti.* See George Sarton, "Ptolemy and His Time," in *Ancient Science and Modern Civilization,* pp. 43 and 64f.; José María López Piñero, *El arte de navegar en la España del Renacimiento,* p. 30.

11. López Piñero, *El arte,* 31.

12. David Romano, *La ciencia hispanojudía,* p. 178ff.; Juan Vernet Gines, *Historia de la ciencia española,* p. 93ff.; López Piñero, *El arte,* pp. 34ff. and 122ff.; and González, *Astronomía,* p. 25ff. Romano dates Zacuto's death to 1522.

13. López Piñero, *El arte,* p. 122; and González, *Astronomía,* p. 33.

14. For more on Portuguese expansion and navigation, see John Law, "On the Methods of Long-Distance Control: Vessels, Navigation, and the Portuguese Route to India," *Sociological Review Monograph* 32 (1986): 234–263.

15. Manuel Sellés, *Instrumentos de navegación: Del Mediterráneo al Pacífico,* p. 43ff.; Patricia Seed, *Ceremonies of Possession in Europe's Conquest of the New World, 1492–1640,* p. 114.

16. Gerard L'E. Turner, *Scientific Instruments, 1500–1900: An Introduction,* p. 30.

17. González, *Astronomía,* p. 31.

18. López Piñero, *El arte,* p. 127.

19. For information on this topic, see Luis de Albuquerque, *Historia de la navegación portuguesa;* Seed, *Ceremonies,* pp. 107–116.

20. Seed, *Ceremonies,* pp. 110–111.

21. Real Cédula de la reina Isabel ordenando el establecimiento de la Casa de la Contratación y Navegación de las Indias, February 14, 1503, Alacalá de Henares, AGI, Contratación 5784, L. 1, ff. 1v–2. In June of 1503 the Casa was moved to the old palace (*alcázar viejo*) of Seville. See Joseph de Veitía Linage, *Norte de la contratación de las Indias occidentales,* pp. 4–5; Goodman, *Power and Penury,* p. 74ff.; Antonio de Herrera y Tordesillas, *Historia general de los hechos de los Castellanos en las Islas i Tierra Firme del Mar Oceano escrita por Antonio de Herrera coronista [sic] major de Su Magestad de las Indias y su coronista de Castilla, en quatro decadas desde el año de 1492 hasta el de 1531,* década 1, p. 144; J. H. Parry, *The Spanish Seaborne Empire,* p. 54f.; Haring, *Trade and Navigation,* Chapter 2.

22. Mariano Cuesta, "El tratado de Tordesillas y la cartografía en la época de los reyes católicos," in *El tratado de Tordesillas en la cartografía histórica,* ed. Jesús Varela-Marcos, p. 84, note 37.

23. D. W. Waters, *The Art of Navigation,* p. 62; For more on the Portuguese Casa, see also Law, "On the Methods of Long-Distance Control." The Casa da Guiné gave way to the Casa da India, which was replaced (1604) by the Conselho da India e Conquistas Ultramarinas. See A. J. R. Russell-Wood, *The Portuguese Empire, 1415–1808,* p. 64.

24. On the Casa da India, see Luís de Albuquerque and Francisco Contente Domingues, *Dicionário de história dos descobrimentos portugueses,* "India, Casa da"; and Joel Serrão, ed., *Dicionário de história de Portugal,* "India, Casa da."

25. There is a long literature on the Casa administrative functions; see Veitía Linage, *Norte;* Haring, *Trade and Navigation,* Chapters 2 and 12; and José Manuel Piernas Hurtado, *La Casa de la Contratación de las Indias.* For information on the particular scientific activities of the Casa, see José Pulido Rubio, *El piloto mayor de la Casa de la Contratación de Sevilla: Pilotos mayores, catedráticos de cosmografía y cosmógrafos;* Manuel de la Puente y Olea, *Los trabajos geográficos de la Casa de Contratación.*

26. Pulido, *El piloto,* p. 18, note 2.

27. See Veitía Linage, *Norte,* pp. 611 (chief pilot; the correct date is 1508), 246f. (ship's inspector), 619 (cosmographer for the making of instruments; for this date see note 34 below), 618 (cosmographer for the chair on cosmography), 258 (representative in Cádiz), 32 (president), 61 (lawyers), 408 (chamber of justice), 171 (treasurer of *haberías*), 227 (supplier), and 132 (bailiff). Veitía Linage claims that the office of the president was established in 1557, but Schäfer corrects him; see Schäfer, *El Consejo Real,* vol. 1, p. 376, note 2. For an English account of the Casa activities and structure, see Haring, *Trade and Navigation,* Chapter 2.

28. Veitía Linage, *Norte,* p. 4ff.

29. See Ordenanzas de la Casa de la Contratación de 1552 in *Colección Navarrete,* vol. 3, p. 197.

30. See the decree of 1503, AGI, Contratación 5784, L. 1, ff. 1v–2; see also Veitía Linage, *Norte;* Carla Rahn Phillips, *Six Galleons for the King of Spain: Imperial Defense in the Early Sixteenth Century,* p. 9.

31. Real Cédula nombrando a Americo Vespuccio como piloto mayor de la Casa de la Contratación, March 22, 1508, Burgos, AGI, Contratación 5784, L. 1, f. 4v; and Real Provisión a Americo Vespuccio, August 6, 1508, Valladolid, AGI, Indiferente 1961, L. 1, ff. 65v–67.

32. See also Paula de Vos's argument on the institutionalization of science at the Casa de la Contratación: "Spain in the 'New World' of Europe: The Casa de la Contratación and the Organization of Navigation and Cartography in the Sixteenth Century." I am grateful to de Vos for making her thesis available to me. For information on the institutionalization of science in France, see Maurice Crosland, "The Development of a Professional Career in Science in France," in *The Emergence of Science in Western Europe,* ed. Maurice Crosland, pp. 127–159.

33. This was the first governmental body in charge of the general administration of the Indies. With Charles, this governmental body was replaced by a group of royal officials that would constitute the Council of Indies in 1524; see Schäfer, *El Consejo Real,* vol. 1, pp. 24–55; and Schäfer, "El origen."

34. At the death of Queen Isabella, the crown of Castille passed to Philip I for less than a year, until his death in September 1506; Ferdinand left Spain during that time but came back as king after Philip's death. For the quotation, see Herrera, *Historia general,* década I, libro 7, p. 177; this translation is from Antonio de Herrera y Tordesillas, *The General History of the Vast Continent and Islands of America, Commonly call'd, The West-Indies, From the First Discovery thereof: With the Best Accounts the People could give of their Antiquities. Collected from the Original Relations sent to the Kings of Spain. By Antonio de Herrera, Historiographer of His Catholic Majesty. Translated into English by Capt. John Stevens,* vol. 1, pp. 321–322.

35. Amerigo Vespucci was appointed chief pilot on March 22, 1508; see Real Cédula a los oficiales de la Casa de la Contratación, March 22, 1508, Burgos, AGI, Contratación 5784, L. 1, f. 4v.

36. On the name "America," see Dietrich Briesemeister, "La imagen de América en la Alemania que conoció Hernando Colón," in *Hernando Colón y su época,* p. 31.

37. Real Provisión a Juan Díaz de Solís y Juan Vespucci, July 24, 1512, Burgos, AGI, Contratación 5784, L. 1, ff. 20r–21r: "Por quanto a nuestra noticia es venydo y por esperiencia se ha visto que por no ser los pilotos tan espertos ni tan instrutos como seria menester para regir y governar los navios que llevan a cargo en los viajes que hazen para las yndias y slas y tierra firme del mar Oceano y por defecto dellos por no saber de que maña se han de regir y governar ny por donde han de tomar el quadrante y el astrolabio y el altura ny saber la quarta della les han acaecido y en cada dia acaecen muchos yerros y defectos en las navegaciones que hazen de lo qual a nos se . . . mucho deservicio y a los tratantes en las dichas yndias mucho daño y de cada dia se espera recibyr mayor [daño] sy no lo mandamos probeer y remediar."

38. Provisión Real a Americo Vespuccio, August 6, 1508, Valladolid, AGI, Indiferente 1961, L. 1, ff. 65v–67. See also Veitía Linage, *Norte,* p. 610.

39. Provisión Real a Juan Díaz de Solís y Juan Vespucci, July 24, 1512, Burgos, AGI, Contratación 5784, L. 1, ff. 20r–21r.

40. This incident is recorded in Richard Eden, *The Decades of the Newe Worlde or West India, Contynyng the Navigations and Conquestes of the Spanyardes, with the Particular Description of the Moste Ryche and Large Landes and Ilandes Lately Founde in the West Ocean Perteynyng to the Inheritaunce of the Kinges of Spayne. In the Which the Diligente Reader May Not Only Consyder What Commoditie May Heraby Chaunce to the Hole Chirstian World in Tyme to Come, But Also Learne Many Secreates Touchynge the Lande, the Sea, and the Starres, Very Necessarie to Be Knowen to Al Such as Shal Attempte Any Navigationes, or Otherwise Have Delite to Beholde the Strange and Wooderfull Woorkes of God and Nature. Written in the Latin Tonge by Peter Martyr of Angleria, and Translated into Englishe by Rycharde Eden,* p. 244 verso.

41. Real Provisión a Americo Vespuccio, August 6, 1508, Valladolid, AGI, Indiferente 1961, L. 1, ff. 65v–67: "visto que por no ser los pilotos tan expertos como sería menester ni tan instruidos en lo que deban saber que baste para regir y gobernar los navios que navegan en los viajes que se hacen por el mar oceano a las nuestras islas y Tierra Firme que tenemos en la parte de las Indias e por defecto dellos e de no saber como se han de regir e governar [las naos] e de no tener fundamento para saber viajar por el cuadrante, e astrolabio el altura ni saber la cuenta dello, les han acaecido muchos yerros, e las gentes que de bajo de su gobernanción navegan han pasado muchos peligros, de que Nuestro Señor ha sido deservido, e en nuestra hacienda, e de los mercaderes que hallan contratan se ha recibido mucho daño e perdida e por remediar lo susodicho e porque es necesario que asi para la dicha navegación, como para otras navegaciones que con ayuda de Nuestro Señor esperamos mandar hacer para descubrir otras tierras es necesario que haya personas mas expertas y mejor fundadas que sepan las cosas necesarias para las tales navegaciones e los que bajo de ellos puedan ir mas seguramente, es nuestra merced e voluntad e mandamos a todos los pilotos de nuestros reinos y señorios que ahora son y seran de aqui adelante que quisieren ir por pilotos en la dicha navegacion de las dicha islas e tierra firme que tenemos a las partes de las Indias e otras partes del mar oceano sean instruidos y sepan lo que es necesario de saber en el cuadrante y astrolabio, para que junta la platica con la teoria se puedan aprovechar de ello en los viajes que hiciera a las dichas partes y que sin lo saber no puedan ir en los dichos navios por pilotos ni en ganar soldadas por pilotaje." Therefore, "mandamos [los reyes a Vespucci] que les enseñeis [a los pilotos el uso de esos instrumentos] en vuestra casa, en Sevilla, a todos los que lo quisieren saber, pagandoos vuestro trabajo." See a partial transcription of this royal decree in Pulido, *El piloto,* pp. 66–67.

42. For information on Cabot's life, see José Toribio Medina, *El Veneciano Sebastián Caboto al servicio de España y especialmente de su proyectado viaje a las Molucas por el estrecho de Magallanes y al reconocimiento de la costa del continente hasta la gobernación de Pedrarias Dávila;* and Robert W. Karrow, *Mapmakers of the Sixteenth Century and Their Maps,* pp. 103–112.

43. Karrow, *Mapmakers,* pp. 104–105.

44. El rey Fernando a Milord de Vliby [*sic*], September 13, 1512, Logroño, AGI Indiferente 419, L. 4, ff. 19r and 19v; see also Medina, *El Veneciano,* vol. 1, p. 2.

45. El rey Fernando a los Oficiales de la Casa de la Contratación, October 22, 1512, Logroño, in Medina, *El Veneciano,* vol. 1, p. 4.

46. Haring, *Trade and Navigation,* p. 37.

47. Testimonio de una Real Cédula por la que se ordena el método que Caboto debe observar en el examen de pilotos, August 2, 1527, AGI, Patronato 251, R. 22.

48. For a description of Seville in this year, see Andrea Navagero, *Viaje a España del magnífico señor Andrés Navagero (1524–1526) embajador de la República de Venecia ante el Emperador Carlos V.*

49. For information on the role of the Englishmen in this expedition, see Boies Penrose, *Tudor and Early Stuart Voyaging.*

50. Karrow, *Mapmakers,* p. 107.

51. Ordenanzas para la Casa de la Contratación elaboradas por el Obispo de Lugo, Juan Suárez de Carvajal del Consejo de su magestad, December 9, 1536, Sevilla, AGI, Patronato 251, R. 33.

52. Certificación de un de lo capítulo sobre el examen de pilotos de las ordenanzas hechas por el licenciado Gregorio López, April 29, 1546, Seville, AGI, Patronato 2590 R. 14.

53. Ordenanzas de 1552 in *Colección Navarrete,* vol. 3, p. 197.

54. Pleito entre Sancho Gutiérrez y Alonso de Chaves, 1556–1561, AGI, Justicia 768, No. 2; see also Pulido, *El piloto,* pp. 137–138; *Colección Navarrete,* vol. 3, p. 197.

55. Veitía Linage, *Norte,* p. 619; Pulido, *El piloto,* p. 292; Armando Cortesão, *Cartografia e cartógrafos portugueses dos séculos XV e XVI,* vol. 2, p. 130ff.

56. The expedition to the Molucca Islands was originally assigned to Fernando de Magellan and Ruy Faleiro. See the Capitulación tomada con Magallanes y Faleiro on March 22, 1518, in *Colección Navarrete,* vol. 16, pp. 131–140, and additional documents on p. 115ff.

57. See Mariano Cuevas, *Monje y marino: La vida de fray Andrés de Urdaneta,* p. 50.

58. Cortesão, *Cartografia,* vol. 1, p. 256. Cortesão discusses Herrera's account about Faleiro (as well as Pedro and Jorge Reinel) and corrects him with documentary evidence; see Herrera, *Historia general,* década 3, libro 4, cap. 13, p. 132.

59. For information on Pedro Reinel (father) and Jorge Reinel (son), see Cortesão, *Cartografia,* vol. 1, p. 249ff.

60. See José M. López Piñero et al., eds., *Diccionario histórico de la ciencia moderna en España,* p. 225f.; see also Germán Latorre, *Diego Ribero: Cosmógrafo y cartógrafo de la Casa de la Contratación de Sevilla.*

61. Cortesão, *Cartografia,* vol. 1, p. 253. Medina says that he made only eighteen: *El Veneciano,* vol. 1, p. 322. Cortesão's sources seem more reliable in this case.

62. For the appointment of Nuño García, see Haring, *Trade and Navigation,* p. 36; Reinel was appointed with a salary of 30,000 maravedís on May 27, 1524; see Real Cédula nombrando a Jorge Reinel, May 24, 1524, Burgos, AGI, Indiferente 1204, No. 21 (3) and (4).

63. Ursula Lamb was the historian who first studied the negotiating and judicial practices of the Spanish pilots and the Casa meetings. More important, she argued that there was a relationship between those practices and science. In this regard my work follows her lead. See Ursula Lamb, "Science by Litigation: A Cosmographic Feud," *Terrae Incognitae* 1 (1969): 40–57; Ursula Lamb, "The Spanish Cosmographic Juntas of the Sixteenth Century," *Terrae Incognitae* 6 (1974): 51–64; Ursula Lamb, "Cosmographers of Seville: Nautical Science and Social Experience," in *First Images of America: The Impact of the New World on the Old,* ed. Fredi Chiappelli, pp. 675–686; Karrow, *Mapmakers,* p. 285; Medina, *El Veneciano,* vol. 1, p. 355; see also the more recent work by Alison Sandman, "Mirroring the World: Sea Charts, Navigations, and Territorial Claims in Sixteenth-Century Spain," in *Merchants and Marvels,* ed. Smith and Findlen.

64. Carta de Fray Juan Caro a su cuñado, Dr. Porras, en Sevilla, December 19,

1525, Conchin, India, in *Colección Navarrete,* vol. 16, pp. 278–280; Carta de Fray Juan Caro al Emperador, December 29, 1526, Cochin, India, in *Colección Navarrete,* vol. 16, pp. 280–284. Caro's significance is related to the Moluccas dispute rather than to the evolution of the teaching practices at the Casa; his proposal is significant, however, because it shows, once again, the degree to which "commoners" participated in proposing institutional novelties to the crown. I was unable to find Caro's date of death. See Cortesão, *Cartografia,* vol. 2, pp. 20f. and 416f.

65. Real Cédula a Fernando Colón, June 13, 1523, Valladolid, in Rodolfo del Castillo Quartiellerz, *Documento inédito del siglo XVI referente a D. Fernando Colón,* p. 8ff.

66. Juan Pérez, "Memoria de las obras y libros de Hernando Colón," in *"Memoria de las obras y libros de Hernando Colón" del bachiller Juan Pérez,* ed. Tomás Marín Martínez, p. 47. Colón's unfinished manuscript is housed in the Biblioteca Colombina in Seville. See Richard L. Kagan, "Clio and the Crown: Writing History in Habsburg Spain," in *Spain, Europe and the Atlantic World: Essays in Honour of John H. Elliott,* ed. Richard L. Kagan and Geoffrey Parker, p. 86.

67. Real Cédula a Hernando Colón, June 13, 1523, Valladolid, in Castillo Quartiellerz, *Documento inédito,* p. 8ff.

68. Antonio Rumeu de Armas, *Hernando Colón, historiador del descubrimiento de América,* p. 76f. See the reports of the Junta in *Colección Navarrete,* vol. 16, p. 745ff.; Herrera, *Historia general,* década 3, libro 6, cap. 6, pp. 183–184.

69. Real Cédula a Hernando Colón para que termine la carta de navegación que se le ordenó en cédula real de octubre 6 de 1526 (Granada), May 5, 1535, Madrid, AGI, Indiferente 1961, L. 3, ff. 276r–276v.

70. Real Provisión a Alonso de Chaves, August 21, 1528, Madrid, AGI, Indiferente 421, L. 13, f. 295v.

71. The statutes of 1552 are in *Colección Navarrete,* vol. 3, pp. 105–263.

72. On the life of Jerónimo de Chaves, I am following Karrow, *Mapmakers,* pp. 116–117.

73. Jerónimo de Chaves, *Tratado de la Sphera que compuso el doctor Ioannes de Sacrobusto con muchas adiciones. Agora nuevamente traduzido de Latin en lengua Castellana por el bachiller Hierónymo de Chaves: el qual añidio muchas figuras, tablas, y claras demostraciones: juntamente con unos breves scholios, necessarios a mayor illucidation, ornato y perfection del dicho tratado.*

74. Jerónimo de Chaves, *Chronographia o Repertorio de los tiempos, el mas copioso y preciso que hasta agora ha salido a luz; en el qual se tocan y declaran materias muy provechosas de philosophia, astrologia, cosmographia y medicina. . . .*

75. Karrow, *Mapmakers,* pp. 116–117; Antonio Palau y Dulcet, *Manual del librero hispano-americano,* No. 67450.

76. Nombramiento de Jerónimo de Chaves, December 4, 1552, Monzón, AGI, Contratación 5784, L. 1, ff. 95–95v.

77. Ibid.

78. *Colección Navarrete,* vol. 3, p. 197ff.

79. In general, the lecturer of the chair of cosmography also made instruments. See Pulido, *El piloto,* p. 25.

80. See the Real Provisión a Americo Vespuccio, August 6, 1508, Valladolid, AGI, Indiferente 1961, L. 1, ff. 65v–67.

81. Not to be confused with the physician Francisco Hernández.

82. Carta al rey del Dr. Hernández, September 22, 1549, AGI, Indiferente 1093, R. 5, No. 98.

83. Memorial de los pilotos, August 20, 1578, Sevilla, AGI, Indiferente 1095, R. 1, No. 20-B/3.

84. Rodrigo Zamorano was appointed to the chair of cosmography on November 20, 1575; cosmographer of the Casa de la Contratación on August 26, 1579; and finally chief pilot on April 13, 1586. See Pulido, *El piloto,* p. 26.

85. Juan de Escalante de Mendoza, *Itinerario de navegación de los mares y tierras occidentales.* Escalante finished his book in 1575.

86. For information on Juan de Herrera and his scientific activities, see Catherine Wilkinson-Zerner, *Juan de Herrera: Architect to Philip II of Spain;* Luis Cervera Vera, *Estudios sobre Juan de Herrera;* and Francisco Javier Sánchez Cantón, *La librería de Juan de Herrera.*

87. Alonso Alvarez de Toledo personally received the instruments: Conocimiento de los instrumentos que recibió Alonso Alvarez de Toledo, January 8, 1574, Madrid, AGI, Patronato 259, R. 58. For a full description of the instruments, see this document.

88. Memorial de Alonso Alvarez de Toledo, Capítulos 1, 3, and 11, August 20, 1578, Sevilla, AGI, Indiferente 1095, R. 1, No. 20-B/4; *Recopilación de leyes de los reynos de las Indias,* vol. 3, libro 9, título 23, ley 38, statute 183, folio 291: "El piloto y maestre en cada puerto donde llegaren, tomen la altura del Sol, ante el escrivano del Navio: y asi mismo pongan baxos e islas, que de nuevo se descubrieren, y no estuvieren en las cartas, y lo entreguen todo por testimonio ante el presidente, y jueces de la Casa [de la Contratación]."

89. Memorial de Alonso Alvarez de Toledo, capítulos 2 and 4, August 20, 1578, Sevilla, AGI, Indiferente 1095, R. 1, No. 20-B/4.

90. Ibid., capítulos 5 and 6.

91. Ibid., capítulos 8, 9, and 10.

92. Ibid., capítulo 7.

93. Carta del licenciado Gamboa, August 20, 1578, Sevilla, AGI, Indiferente 1095, R. 1, No. 20-B/1.

94. Memorial de los pilotos, August 20, 1578, Sevilla, AGI, Indiferente 1095, R. 1, No. 20-B/3.

95. Memorial del piloto mayor, Alonso de Chaves, August 20, 1578, Sevilla, AGI, Indiferente 1095, R. 1, No. 20-B/6.

96. Memorial del capitán Juan Escalante de Mendoza, August 20, 1578, Sevilla, AGI, Indiferente 1095, R. 1, No. 20-B/5.

97. Memorial de Rodrigo Zamorano, August 20, 1578, Sevilla, AGI, Indiferente 1095, R. 1, No. 20-B/2.

98. Memoriales de los pilotos y del piloto mayor, Alonso de Chaves, August 20, 1578, Sevilla, AGI, Indiferente 1095, R. 1, No. 20-B/3 and B/6.

99. Memorial del capitán Juan Escalante de Mendoza, August 20, 1578, Sevilla, AGI, Indiferente 1095, R. 1, No. 20-B/5.

100. Expedientes, informaciones y probanzas, 1573, AGI, Indiferente 1223, cited in Paulino Castañeda Delgado, Mariano Cuesta Domingo, and Pilar Hernández Aparicio, "Estudio preliminar," in *Alonso de Chaves: Quatri partitu en cosmografía práctica, y por otro nombre Espejo de navegantes,* p. 31.

101. Memorial de Rodrigo Zamorano, August 20, 1578, Sevilla, AGI, Indiferente 1095, R. 1, No. 20-B/2.

102. Memorial del piloto mayor, Alonso de Chaves, August 20, 1578, Sevilla, AGI, Indiferente 1095, R. 1, No. 20-B/6.

103. Memorial de los pilotos, August 20, 1578, Sevilla, AGI, Indiferente 1095, R. 1, No. 20-B/3.

104. Memorial del capitán Juan Escalante de Mendoza, August 20, 1578, Sevilla, AGI, Indiferente 1095, R. 1, No. 20-B/5.

105. Memorial de Rodrigo Zamorano, August 20, 1578, Sevilla, AGI, Indiferente 1095, R. 1, No. 20-B/2.

106. Memorial de los pilotos, August 20, 1578, Sevilla, AGI, Indiferente 1095, R. 1, No. 20-B/3.

107. Memorial del capitán Juan Escalante de Mendoza, August 20, 1578, Sevilla, AGI, Indiferente 1095, R. 1, No. 20-B/5.

108. Ibid., R. 1, No. 20-B/5.

3. COMMUNITIES OF EXPERTS

1. Miguel León Portilla, ed., *The Broken Spears: The Aztec Account of the Conquest of Mexico,* p. 16.

2. For more on this topic, see Parry, *Discovery,* p. 149ff.; Seed, *Ceremonies;* for more on the development of ships and guns, see Carlo M. Cipolla, *Guns, Sails, and Empires: Technological Innovation and the Early Phases of European Expansion, 1400–1700,* p. 75ff.

3. See Bruno Latour, *Science in Action: How to Follow Scientists and Engineers through Society,* Chapter 1.

4. For more on the idea of progress and technology, see López Piñero, *Ciencia y técnica,* pp. 157–159; and Isabel Arenas Frutos, "Inventos sobre tecnología submarina para la América colonial," *Asociación de Historiadores Latinoamericanistas Europeos (AHILA),* special issue (1992): 421–434.

5. For information on the relations between practice and theory in the Middle Ages, see Pamela O. Long, "Science and Technology in Medieval Society," *Annals of the New York Academy of Sciences* 441 (1985); Jean Gimpel, *The Medieval Machine;* and Lynn White, *Medieval Technology and Social Change.* For the later period, see Paolo Rossi, *Philosophy, Technology, and the Arts in the Early Modern Era,* trans. Salvator Attanasio; and Pamela H. Smith, *The Business of Alchemy.*

6. I am following Nicolás García Tapia, *Técnica y poder en Castilla durante los siglos XVI y XVII.*

7. Real Cédula a los oficiales de la Casa de la Contratación, December 30, 1537, Valladolid, AGI, Indiferente 1962, L. 5, ff. 309v–310; On the German mining-masters, see Pedro Simón, *Noticias historiales de las conquistas de Tierra Firme en las Indias Occidentales,* vol. 1, p. 183.

8. García Tapia, *Técnica,* p. 30ff.

9. Ibid., p. 49ff.

10. Goodman, *Power and Penury,* p. 66.

11. AGI, several documents on patents; see also García Tapia, *Técnica,* p. 195.

12. Pablo Emilio Pérez-Mallaína Bueno establishes similar categories for the study of instruments; see his *Los inventos llevados de España a las Indias en la segunda mitad del siglo XVI,* p. 36; Manuel Luengo Muñoz, "Inventos para acrecentar la obtención de perlas en América, durante el siglo XVI," *Anuario de Estudios Americanos* 9 (1952): 51–72; Arenas Frutos, "Inventos"; Enrique Otte, "El proceso del rastro de perlas de Luis de Lampiñan," *Boletín de la Academia Nacional de la Historia de Caracas* 187 (1964): 386–406; García Tapia, *Técnica.*

13. González, *Astronomía,* p. 22.

14. Otte, "El proceso," p. 386.

15. Real Cédula licencia a Juan de Cárdenas, August 27, 1520, AGI, Indiferente 420, L. 8, ff. 253v–255r; see also Luengo Muñoz, "Inventos," pp. 54–55.

16. Real Cédula licencia a Juan de Cárdenas, ff. 253v–255r.

17. See David Brading, *The First America: The Spanish Monarchy, Creole Patriots, and the Liberal State, 1492–1867.*

18. Cortesão, *Cartografia,* vol. 2, p. 137.

19. Relación de Diego Ribeiro, n.d., AGI, Indiferente 1528, No. 29. The document is perhaps from 1524, because the royal answer to this report is dated 1524; see Latorre, *Ribero,* p. 11ff.

20. See the Real Cédula of March 11, 1532, Ocaña, AGI, Indiferente 422, L. 15, ff. 18v–19r.

21. Real Cédula otorgada a Diego Ribeiro, November 26, 1526, Granada, in Latorre, *Ribero,* pp. 17.

22. Real Cédula a los oficiales de la Casa de la Contratación, July 17, 1531, Avila, AGI, Indiferente 1961, L. 2, ff. 84v–85; and Real Cédula a los oficiales de la Casa de la Contratación, October 13, 1531, Medina del Campo, AGI, Indiferente 1961, L. 2, f. 99v.

23. Real Cédula a los oficiales de la Casa de la Contratación, November 4, 1531, Medina del Campo, AGI, Indiferente 1961, L. 2, ff. 105v–106.

24. See Latorre, *Ribero,* p. 18ff.

25. Cortesão, *Cartografía,* vol. 2, p. 137.

26. Consulta del Consejo, May 4, 1570, Madrid, AGI, Indiferente 738, No. 119. The council made that statement in the context of the mining invention made by Benito de Morales.

27. See Latorre, *Ribero,* p. 16.

28. Carta del Consejo de Indias a los oficiales de la Casa de la Contratación, June 28, 1535, Madrid, AGI, Indiferente 1961, L. 3, ff. 293v–294.

29. Cédula Real a Vicente Barreros, January 22, 1536, Madrid, AGI, Indiferente 1962, L. 4, ff. 31v–32v.

30. On the disciplining of experience for the English case, see Steven Shapin, "The House of Experiment in Seventeenth-Century England," *Isis* 79 (1988): 375.

31. Real Cédula a Johan Fernández de Castro, June 20, 1526, Granada, AGI, Indiferente 421, L. 11, ff. 41r–42r.

32. Castro was involved in other business with the crown. In 1529, for instance, he obtained a contract to trade the "wood of Brazil"; see AGI, Patronato 170, No. 36. Brazilwood was used to extract a red dye.

33. Licencia otorgada a Johan Fernández de Castro, June 20, 1526, Granada, AGI, Indiferente 421, L. 11, ff. 41r–42r.

34. Martín Cortés, *Breve compendio de la sphera y de la arte de navegar, con nuevos instrumentos y reglas, ejemplificado con muy subtiles demostraciones,* f. ii verso.

35. For information on mining, see Modesto Bargalló, *La minería y la metalurgia en la América española durante la época colonial;* González Loscertales, Vicente Roldán de Montaud, and Inés Roldán de Montaud, "La minería del cobre en Cuba, su organización, problemas administrativos y repercusiones sociales (1828–1849)," *Revista de Indias* 40 (1980): 255–299; Frédérique Langue and Carmen Salazar Solero, *Dictionnaire des termes miniers en usage en Amérique espagnole (XVIe–XIXe siècle)/Diccionario de términos mineros para la América española (siglos XVI a XIX);* Guillermo Lohmann Villena, "La minería en el marco del Virreinato peruano," in *I Coloquio Internacional sobre Historia de la Minería,* pp. 639–665; and Demetrio Ramos Pérez, "Ordenación de la minería en Hispanoamérica durante la época provincial (siglos XVI, XVII, XVIII)," in *I Coloquio Internacional sobre Historia de la Minería,* pp. 373–397; Rosario Sevilla Soler, "La minería americana y la crisis del siglo XVII: Estado del problema," *Anuario de Estudios Americanos,* special supplement 47, no. 2 (1990): 61–81; Luis Muro, "Bartolomé de Medina, introductor del beneficio de patio en Nueva España," *Historia Mexicana* 13, no. 4 (1964): 517–531; Peter Bakewell, "Technological Change in Potosí: The Silver Boom of the 1570's," *Jahrbuch für Geschichte von Staat, Wirtschaft und Gesellschaft Lateinamerikas* 14 (1977): 57–77; and Julio Sánchez Gómez, "La técnica en la producción de metales monedables en España y en América, 1550–1650," in *La savia del imperio: Tres estudios de economía colonial,* ed. Julio Sánchez Gómez et al. I am very grateful to Professor Julio Sánchez for his comments on this subject.

36. Sánchez Gómez, "La técnica," p. 73ff.

37. Capitulación entre el Príncipe Felipe y Juan Tetzel, January 11, 1546, Madrid, AGI, Santo Domingo, 99, R. 6, No. 22. The historian Demetrio Ramos Pérez argues that the legislation concerning mines in the New World was modified to suit the purposes of the German bankers; see Ramos Pérez, "Ordenación de la minería," p. 381.

38. Capitulación entre el Príncipe Felipe y Juan Tetzel.

39. Concierto entre Juan Tetzel y el cabildo de la ciudad de Santiago, isla Fernandina, June, 27, 1550, AGI, Santo Domingo, 99, R. 6, No. 22; another copy in Patronato

238, No. 2, R. 1.: The council says that "hasta ahora [Teçel] no ha querido manifestar el secreto de la manera que se ha de tener en lo fundir [el cobre] y poner en estado que se pueda muy buenamente labrar [las minas] y los vecinos y moradores desta isla an pretendido y pretenden que el dicho Juan Teçel es obligado a manifestar el secreto de la condición y aprovechamiento del fundir del dicho cobre para que ellos puedan gozar del beneficio de las dichas minas y que [esa fue] la intención de su mag."

40. Concierto entre Juan Tetzel y el cabildo.

41. Carta de Juan Tetzel al rey, March 15, 1571 [this is the date of the reply; the actual letter is not dated], Madrid, AGI, Santo Domingo, 99, R. 6, No. 22.

42. Ibid.

43. See these documents: Carta de Juan Tetzel al rey, March 15, 1571, Madrid, AGI, Santo Domingo, 99, R. 6, No. 22; and Capitulación entre el rey y Sancho Medina, January 20, 1572, Madrid, AGI, Patronato 238, No. 2, R. 1. In this document Sancho de Medina mentions that Juan Tetzel has died.

44. In all the documents the names of these foreigners appear in their Spanish form. In 1536 Juan Enchel came with other Germans to New Spain to work at the silver mines. They brought and built devices and mills to work at the mines; see Francisco A. de Icaza, *Diccionario autobiográfico de conquistadores y pobladores de Nueva España,* vol. 2, p. 258, note 1156. In 1592 Miguel Redelic or Miguel Alemán had been living in the New World for twenty-eight years. In 1592 he was working in the mines of San Andrés, New Mexico; see "Proceso de la Inquisición contra Miguel Redelic, alias Miguel Alemán," Mexico City, November 8, 1592, Bancroft Library, University of California–Berkeley.

45. Sánchez Gómez, "La técnica," part 2; Germans also participated in the expeditionary force sent by viceroy Don Antonio de Mendoza to help *licenciado* Pedro de la Gasca in Peru; see Cuevas, *Monje y marino,* p. 138.

46. Henry Kamen, *Philip of Spain,* p. 182; Francisco Iñiguez Almech, *Casas reales y jardines de Felipe II,* pp. 125, 156, 177, 201ff.

47. Bargalló, *La minería,* p. 56f.

48. "Petición de la ciudad de México sobre el repartimiento general y perpetuo de la Nueva España: La presentó Juan Velázquez de Salazar al licenciado Juan de Ovando en Madrid, a 6 de junio de 1571," in *Epistolario de Nueva España: 1505–1818,* ed. Francisco del Paso y Troncoso, vol. 11, Doc. 659, p. 118; Bargalló, *La minería,* p. 91.

49. Muro, "Bartolomé de Medina," p. 524.

50. Paso y Troncoso, *Epistolario,* vol. 11, Doc. 659, p. 118.

51. Juan Manuel Menes Llaguno, *Bartolomé de Medina: Un sevillano pachuquero,* p. 47ff.; Manuel Castillo Martos and Mervyn Francis Lang, *Metales preciosos: Unión de dos mundos,* p. 96.

52. Merced, undated [ca. November 1554?], Mexico City, Archivo General de la Nación, Mexico City, Ramo de Mercedes, V, ff. 87–87v, transcribed in Muro, "Bartolomé de Medina," p. 518ff.: Bartolomé de Medina, before the viceroy Don Luis de Velasco, said "que estando en España él tuvo noticia de la orden que tenía en esta tierra en el beneficiar los metales de oro y plata y las grandes costas y repartos que en ello había, y para saber si era así había pasado a esta Nueva España a lo ver por vista de

ojos y a procurar como los dichos metales se beneficiasen a menos costa y así con gran diligencia y cuidado y trabajo de su persona y costa de su hacienda había entendido por la experiencia que tenía de los susodicho en dar orden como con azogue se pueden beneficiar los dichos metales y se saque dellos toda ley que se le saca por fundición com mucha menos costa de gente y caballos y sin greta y çendrada, carbón ni leña, de lo cual se seguirá gran pro[vecho?] en general a toda esta tierra y acrecentimiento de las rentas reales."

53. For more on this point, see Castillo Martos and Lang, *Metales,* p. 97.

54. Nothing is known about this German expert save that Medina tried to go with him to New Spain, but he did not obtain a license. Medina went alone and worked for almost a year in the new method. See Castillo Martos and Lang, *Metales,* p. 97f.; Eli de Gortari, *La ciencia en la historia de México,* p. 198f.

55. Don Francisco de Mendoza, administrator of the Guadalcanal mines, in his instructions of October 30, 1557, was ordered to implement the amalgamation process in those mines. After some unsuccessful experiments, the *indiano* Mosén Boteller was called from New Spain to Guadalcanal to implement the new method; see Bargalló, *La minería,* pp. 121 and 111.

56. Castillo Martos and Lang, *Metales,* p. 99ff.

57. Biringucci described the process in his *De la Pirotechnia* (1540), Book 9, Chapter 11; see Bargalló, *La minería,* p. 107ff. See the English translation by Cyril Stanley Smith and Martha Teach Gnudi, *The Pirotechnia of Vannocio Biringucci: The Classic Sixteenth-Century Treatise on Metals and Metallurgy.*

58. Muro, "Bartolomé de Medina," p. 522; Bargalló calculated that there were around 126 people using Medina's technology; see Bargalló, *La minería,* p. 112.

59. There is, of course, a theoretical dimension to the Scientific Revolution that is not part of what I call the early Scientific Revolution. The early Scientific Revolution was about the method, the procedures, to transform personal experience into knowledge and this knowledge into practical activities. The story of this aspect of the Scientific Revolution is connected with the changes taking place in Spain and, as a consequence, in the rest of Europe. This empirical method supported the theoretical aspects developed during the seventeenth century. Copernicus, for instance, was a humanist, not an empiricist, and his system became viable only after the empirical evidence gathered by Galileo and others in the seventeenth century.

60. Bakewell, "Technological Change," p. 57.

61. Ibid., p. 58. For more on the *mita,* see Peter Bakewell, "Mining," in *Colonial Spanish America,* ed. Leslie Bethell, pp. 203–249, especially p. 221f.; Guillermo Lohmann Villena, "El Virreinato del Perú," in *El descubrimiento y la fundación de los reinos ultramarinos hasta fines del siglo XVI,* ed. Manuel Lucena Salmoral, pp. 525–562, in particular p. 532ff.

62. For this junta, see Raquel Alvarez Peláez, "La obra de Hernández y su recuperación ilustrada," in *La real expedición botánica a Nueva España, 1787–1803,* p. 147; Alvarez Peláez, *La conquista,* p. 131ff.; Demetrio Ramos Pérez, "La Junta Magna y la nueva política," in *El descubrimiento,* ed. Lucena Samoral, pp. 437–453.

63. Carta del virrey Toledo a Juan de Ovando, June 10, 1572, Cuzco, AGI, Lima 28b, L. 4, ff. 302–304.

64. Bakewell, "Mining," p. 222f.; on the colonial economy of Peru, see Steve J. Stern, *Peru's Indian Peoples and the Challenge of Spanish Conquest,* p. 80ff.

65. Bargalló, *La minería,* p. 130f.; Muro, "Bartolomé de Medina," p. 530, note 30.

66. This account is based on Menes Llaguno and Muro. See Menes Llaguno, *Bartolomé de Medina,* pp. 105–107; and Muro, "Bartolomé de Medina," p. 530, note 30.

67. Carta del virrey don Martín Enríquez, 1576, cited in Bargalló, *La minería,* p. 131f.; Menes Llaguno, *Bartolomé de Medina,* pp. 107–108.

68. Menes Llaguno, *Bartolomé de Medina,* pp. 108–109.

69. Ibid., pp. 105–106; Muro, "Bartolomé de Medina," p. 530, note 30.

70. Menes Llaguno, *Bartolomé de Medina,* p. 106; Muro, "Bartolomé de Medina," p. 530, note 30.

71. Icaza, *Diccionario,* vol. 2, no. 763; Bargalló, *La minería,* p. 94; Menes Llaguno, *Bartolomé de Medina,* p. 107.

72. Muro, "Bartolomé de Medina," p. 530, note 30.

73. The French term was *tartane* (English: tartan). It means a small vessel with a triangular sail. "TARTANE. s. f. Sorte de bastiment de charge, dont on se sert sur la Mer Mediterranée, & qui porte la voile latine": *Dictionnaire de l'Académie française* [online dictionary], 1694 ed., s.v. "Tartane"; available on World Wide Web at http://duras .uchicago.edu /cgi-bin /quick_look.new.sh?word=tartane.

74. Las diligencias de Simón de Bolívar, June 29, 1592, Valladolid, AGI, Escribanía 1008c.

75. Ibid.

76. Ibid.

77. Ibid.

78. Juan de Cárdenas, *Problemas y secretos maravillosos de las Indias,* f. 80r.

79. Ibid., f. 2r.

80. Ibid., f. 86r. The complete text reads: "Lo que en este capitulo se pide es saber, que mysterio o secreto tan estraño aya entre la plata y la sal, que para avella de sacar y desaraygar del propio metal sea forçoso/ hecharle sal, donde no es imposible sacarla, dezir que entre la sal y la plata se halla alguna amistad y conveniencia, y que mediante ella se puede sacar la plata, no lleva esto camino, porque la plata es fria, y humida, la sal caliente y seca, la una de naturaleza de agua, la otra de agua y tierra los effectos de la una y de la otra muy contrarios, segun esto que semajança puede entre los dos aver para mezclar una con otra: pidese pues desto agora la razon, y causa."

81. Ibid., f. 86v. The Spanish text reads: "Toman el metal [los mineros], y muelenlo muy bien, y despues amassanlo con su salmuera, y encorporanlo con tantas libras de azogue, mas o menos segun la ley natural, y ya despues que al cabo d/ algunos dias se presume aver dado el metal la ley (q\ es abraçarse la plata y el azogue) haze el minero lavar el dicho metal, y yendose el barro, y arena del metal con el agua queda como mas pesada en el fondo de la tina aquella massa o pella de plata, y azogue, despues apartan

con fuego el azogue de la plata, y esto se llama sacar plata por azogue, y de estos tales metales es nuestra duda [sobre el rol de la sal]."

82. Ibid., ff. 88r–v.

83. Ibid., ff. 88v–89r. The complete text reads: "Declarado pues, como el azogue por la dicha amistad se abraça con la plata resta agora saber que este acto, y obra de abraçarse, no pudo ser tan breve, veloz, e instantaneo que no uviesse necessidad de tiempo y calor: el tiempo es muy necessario, porque de la misma suerte que los medicamentos tienen necessidad de tiempo cierto, y determinado para poderse encorporar y fermentar, ansi esta massa de plata, metal, y azogue, para poder entre si recozerse, abraçarse y encorporarse, tambien es nescessario el calor para el propio effecto, porque assi como el Ruybarbbe, para atraer assi la colera tiene necessidad d/ actuarse primero, y fortalecerse d/ nuestro calor, assi el azogue tiene necessidad de calor, para aver de abraçarse con la plata. . . ."

84. Ibid., f. 88v. This is the complete text: "La sal se hecha, no para que se abraçe con nadie, sino para que como material caliente sirve de dar calor, y actuar el azogue, y otro si ayudar a recozer, fermentar, y esponjar aq/l metal, porq/ mejor le pueda penetrar el azogue, y abraçarse co/ la plata."

85. Ibid., f. 89v. The text reads: "De donde se infiere que se en lugar de sal se hechase cardenillo, o cal viva, o solimán, o otro material caliente, como no consumiesse, y destryesse el azogue, haria ta/to y mejor effecto que la sal, quanto es material mas caliente."

86. Menes Llaguno, *Bartolomé de Medina,* p. 105; Muro, "Bartolomé de Medina," p. 530, note 30.

87. Cárdenas, *Problemas,* f. 90v. The complete text is as follows: "Supongamos, que un minero tiene un monton de metal de cien quintales, y que a este metal, se hechan para sacarle la plata treynta libras de azogue, todo lo qual se mezcla, incorpora, y repassa hasta venir a tomar la ley: despues haze lavar todo aquel metal, y halla despues de lavado las mesmas treynta libras que hecho primero, no quiero dezir que si hecho de solo azogue treynta libras, essas mesmas halla de azogue, sino que si delas treynta que halla, son las cinco de plata, todo aquello que halla d/ plata, halla menos de azogue. Entra puse agora nuestra dubda, si se sacaron cinco libras de plata, y se hallan menos otras tantas de azogue, que pudo hazerse aquel azogue, y en que se consumio? y quando hallemos causa que al tal azogue aya consumido resta otra mayor difficultad, es a saber porque sea forçoso que precissa y limitadamente se pierda tanto de azogue, quanto se saco de plata, esso pues es lo que al presente se pregunta."

88. Ibid., ff. 91r–92v.

89. Ibid., f. 92v. The complete text: "Para mejor satisfazer nuestra dubda, sera necessario advertir primero, dos admirables propiedades del azogue, delas quales es la primera, ser el dicho azogue compuesto de sutilissimas, y muy penetrativas partes, las quales son tan por estremo delgadas, q/ e/ un pu/to el fuego o otro qualquier calor excessivo las buela, y co/vierte en humo. Esta penetracio/ y subtileza de partes, podemos esperime/tar en alguanas cosas q/ dire, lo primero, bie/ vemos q/ si a uno le untan co/ azogue, le pentra, y passa co/ gran presteza hasta los ppios guessos, lo 2. q/ todas las

vezes q/ pone/ al azogue en co/dicio/ de llegar a fuego, o a color demasiado, al pu/to se buela, y co/vierte en humo. . . ."

90. Ibid., f. 92v. The complete text reads thus: "La segunda propiedad que el azogue tiene, es la mas brava enemistad con el fuego, y con el calor, de quantas se pueden imaginar. Esta enemistad, y contrariedad se hecha de ver palpableme/te, por que el azogue (como despues se dira) es frigidissimo, y muy humido, el fuego caliente, y seco por extremo, el azogue es pesado mas que quantas cosas ay en la naturaleza, y el fuego es ligero, y veloz sobre todas, el azogue es de ser y naturaleza de agua, el fuego es por estremo contrario dellas, ansi que toda la co/trariedad que se halla entre dos enemigos, essa se halla entre el azogue, y el fuego, y por el consiguiente entre el azogue y el calor, porque si el fuego consume el azogue, es mediante el calor."

91. Ibid., f. 93r.

92. Cárdenas uses personal experience not only to discuss the amalgamation process but also to correct Pier Andreas Mattioli's information about maize and the comments of Galen and Hippocrates on the benefits of barley, in contrast to atole (a maize-based drink); see Cárdenas, *Problemas,* ff. 141v and 147.

93. José de Acosta, *Historia natural y moral de las Indias,* Chapter 10.

94. Ibid., p. 241.

95. Consulta del Consejo, May 4, 1570, Madrid, AGI, Indiferente 738, No. 119.

96. Hernández received his instructions in 1570; see AGI, Lima, 569, L. 13, ff. 97v–101r. On Hernández, see Varey, *The Mexican Treasury;* Simon Varey et al., eds., *Searching for the Secrets of Nature: The Life and Works of Dr. Francisco Hernández;* Germán Somolinos D'Ardois, *Vida y obra de Francisco Hernández;* and Germán Somolinos D'Ardois, *El doctor Francisco Hernández y la primera expedición científica en América.*

97. Carta del virrey, December 2, 1572, AGI, Mexico City, 19, No. 97.

98. Expediente sobre el instrumento de Juan Alonso, March 5, 1572, Canaria, AGI, Patronato 264, R. 3.

99. Real Cédula al presidente y oidores de la Audiencia de Quito, August 16, 1572, El Escorial, in Francisco de Solano, ed., *Cuestionarios para la formación de las relaciones geográficas de Indias: Siglos XVI–XIX,* p. 15f.

4. CIRCUITS OF INFORMATION

1. Real Cédula a la Audiencia de México, December 19, 1533, Monzón, transcribed in Solano, *Cuestionarios,* p. 4: "Porque queremos tener entera noticia de las cosas de esa tierra y calidades de ella os mando que luego que ésta recibáis hagáis hacer una muy larga y particular relación de la grandeza de esa tierra, así de ancho como de largo, y de sus límites: poniéndolos muy específicamente por sus nombres propios y cómo se confina y amojana por ellos. Y asimismo, de las calidades y extrañezas que en ella hay, particularmente las de cada pueblo por sí, y qué población de gentes hay en ella de los naturales, poniendo sus ritos y costumbres particularmente."

2. Alvarez Peláez, *La conquista,* p. 131ff.

3. The scholar Marcos Jiménez de la Espada called the reports based on the 1577 questionnaire *Relaciones geográficas de Indias,* and this is the name commonly used today. Raquel Alvarez, however, calls them "Relaciones de Indias" because these reports contained information not only about geography but also about natural history, history, and ethnography. See Jiménez de la Espada, *Relaciones;* and Alvarez Peláez, *La conquista,* p. 15. For more on the relations between geography and state formation, see Chandra Mukerji, *From Graven Images: Patterns of Modern Materialism,* p. 117ff.

4. Max Weber, *The Protestant Ethic and the Spirit of Capitalism,* trans. Talcott Parsons, p. 16.

5. Maravall calls this phase of the early modern Spanish history the "Monarchy against the State." See Maravall, *Estado moderno,* vol. 1, p. 113.

6. Jiménez de la Espada, *Relaciones,* vol. 1, p. 20; Howard F. Cline, "The Relaciones Geográficas of the Spanish Indies, 1577–1648," in *Handbook of Middle American Indians,* ed. Howard F. Cline, vol. 12, part 1, p. 183ff. For more on the encomienda system, see Magnus Mörner, "The Rural Economy and Society of Colonial Spanish South America," in *The Cambridge History of Latin America,* ed. L. Bethell, pp. 190–192; James Lockard, *Spanish Peru: 1534–1560: A Social History,* p. 11ff.; L. F. Calero, *Chiefdoms under Siege: Spain's Rule and Native Adaptation in the Southern Colombian Andes, 1535–1700,* p. 163ff.

7. Jiménez de la Espada, *Relaciones,* vol. 1, p. 33. See also Alvarez Peláez, *La conquista,* p. 81.

8. Capitulación otorgada a Diego de Lepe para ir a descubrir por el Mar Océano, September 14, 1501, Granada, AGI, Indiferente 418, L. 1, ff. 29v–32v; *CODOIN,* series 2, vol. 14, p. 96.

9. Capitulación otorgada a Cristóbal Guerra para ir a descubrir a la Costa de las Perlas y otras islas, July 12, 1503, Alcalá de Henares, in Milagros de Vas Mingo, *Las capitulaciones de Indias en el siglo XVI,* p. 136. The *capitulaciones* with Alonso de Ojeda and Juan de la Cosa also contain references to "monsters"; see Capitulación otorgada a Alonso de Ojeda para ir a descubrir a Coqubacoa, September 30, 1504, Medina del Campo, in Vas Mingo, *Las capitulaciones,* p. 139; and Capitulación otorgada a Juan de la Cosa para ir a descubrir al golfo de Urabá, February 14, 1504, Medina del Campo, in Vas Mingo, *Las capitulaciones,* p. 144.

10. Capitulación otorgada a Diego de Nicuesa y Alonso de Ojeda para comerciar en Urabá y Veragua, June 9, 1508, Burgos, in Vas Mingo, *Las capitulaciones,* p. 156.

11. Real Provisión a Americo Vespuccio, August 6, 1508, Valladolid, AGI, Indiferente 1961, L. 1, ff. 65v–67.

12. Capitulación otorgada a Juan Ponce de León para ir a descubrir y poblar a la isla Bimini, February 23, 1512, Burgos, in Vas Mingo, *Las capitulaciones,* p. 164.

13. Real Provisión a Juan de Solís y Juan Vespucci para hacer el padrón real, July 24, 1512, Burgos, AGI, Contratación 5784, L. 1, ff. 20r–21r; there is a version of this provision in *CODOIN,* series 2, vol. 14, p. 13. See Chapter 2 for a more detailed discussion of this theme.

14. Capitulación otorgada a Diego Velázquez para ir a descubrir y conquistar

Yucatán y Cozumel, November 13, 1518, Zaragoza, in Vas Mingo, *Las capitulaciones,* pp. 169–172.

15. Carta de Alonso de Zauzo al emperador Carlos V, January 22, 1518, La Española, in Jiménez de la Espada, *Relaciones,* vol. 1, p. 11ff.

16. Real Cédula al Licenciado Juan de Cárdenas para rescatar en las islas del Caribe, August 27, 1520, Valladolid, AGI, Indiferente 420, L. 8, ff. 253v–255r.

17. Schäfer, *El Consejo Real,* vol. 1, pp. 24–55.

18. Real Cédula para hacer una carta navegación nueva y verdadera, June 20, 1526, Granada, AGI, Indiferente 421, L. 11, ff. 21v–22v.

19. Real Cédula a Hernando Colón para que termine la carta de navegación que se le ordenó en cédula real de octubre 6 de 1526 (Granada), May 5, 1535, Madrid, AGI, Indiferente 1961, L. 3, ff. 276r–276v.

20. Real Cédula a los pilotos y maestres para que escriban reportes sobre sus viajes, March 16, 1527, AGI, Indiferente 421, quoted in Castañeda Delgado, Cuesta Domingo, and Hernández Aparicio, "Estudio preliminar," p. 20 and note 24. The project of mapping the New World would eventually include the work of indigenous people; see Barbary E. Mundy, *The Mapping of New Spain: Indigenous Cartography and the Maps of the Relaciones Geográficas.*

21. Real Cédula a los oficiales de la audiencia de la isla Española, January 29, 1525, Madrid, AGI, Contratación 5787, No. 1, L. 1, ff. 33r–34v.

22. On Ramírez de Fuenleal's order, see Marcos Jiménez de la Espada, "Antecedentes," in *Relaciones geográficas de Indias;* for more on Gonzalo Fernández de Oviedo, see Chapter 5.

23. Consulta del Consejo, May 27, 1532, Medina del Campo, AGI, Indiferente 737, No. 24. According to Schäfer, this answer was written by Francisco de Cobos, advisor to the emperor; see Schäfer, *El Consejo Real,* vol. 2, p. 405 and note 4.

24. *CODOIN,* series 2, vol. 17, p. 288, August 18, 1532. This document is missing in AGI.

25. Real Cédula al gobernador de la isla Fernandina, October 15, 1532, Segovia, AGI, Indiferente 422, L. 15, ff. 189v–189r. At the end of the decree it says "to send one like this to all the Indies." The royal decree calls him the "captain P[edro] Fernández de Oviedo," but it is the same Gonzalo Fernández de Oviedo of the Consulta of May 27, 1532 (Medina del Campo, AGI, Indiferente 737, No. 24). The wording of the Consulta and the royal decree is the same.

26. Real Cédula al gobernador de la isla Fernandina, October 15, 1532, Segovia, AGI, Indiferente 422, L. 15, ff. 189v–189r.

27. Real Cédula a la Audiencia de la Española, March 3, 1530, Madrid, in Solano, *Cuestionarios,* p. 3.

28. Carta de los oficiales de la Casa de la Contratación, April 13, 1532, Sevilla, AGI, Indiferente 1092, No. 37.

29. Real Cédula a los oficiales de la Casa de la Contratación, November 20, 1532, Valladolid, AGI, Indiferente 1962, L. 5, f. 41v.

30. Real Cédula a don Hernando Colón, May 20, 1535, Madrid, AGI, Indiferente 1961, L. 3, ff. 276r–276v.

31. Carta de Hernán Cortés al Consejo de Indias, n.d., n.p. (before 1533? New Spain?), AHNM, Documentos Diversos 22, doc. 15.

32. Memoria de la derrota y navegación en el descubrimiento del mar del Sur, October 24, 1533–February 16, 1534, Abordo del San Lázaro, AGI, Patronato 20.

33. Ibid., ff. 3v and 4v.

34. Alvarez Peláez, *La conquista*, p. 171f.

35. For information on the reform of the Spanish clergy, see Sara T. Nalle, *God in La Mancha*, p. 95ff.

36. See, for instance, the Real Cédula al Arzobispo de México pidiendo informes sobre las características de la Tierra, November 27, 1548, Valladolid, in Solano, *Cuestionarios*, pp. 5–7; see also Alvarez Peláez, *La conquista*, pp. 173–174.

37. Francisco López de Gómara, *La istoria de las Indias, y conquista de Mexico.*

38. I am following Raquel Alvarez in her description of Santa Cruz's *Memorial;* see Alvarez, *La conquista*, pp. 176–177.

39. The first part of this document is in AGI, Patronato 34, R. 13. I found the second and final part of this document in AGI, Indiferente 857. They are without folio numbers.

40. Santa Cruz, quoted in Geoffrey Parker, "Maps and Ministers: The Spanish Habsburgs," in *Monarchs, Ministers, and Maps*, ed. David Buisseret, p. 126.

41. Ibid., p. 126.

42. Ibid., p. 129ff.; Maravall, *Estado moderno*, vol. 1, p. 203; Richard Kagan, "Philip II and the Art of Cityscape," *Journal of Interdisciplinary History* 17, no. 1 (1986): 120ff. The Spanish scholar Felipe Picatoste y Rodríguez argues that these Spanish projects resulted from projects to map the New World. See Felipe Picatoste y Rodríguez, *Apuntes para una biblioteca científica española del siglo XVI: Estudios biográficos y bibliográficos de ciencias exactas, físicas y naturales y sus inmediatas aplicaciones en dicho siglo*, p. 86ff.

43. Benavides, *Secretos de Chirurgia.*

44. Tomás López Medel, "Tratado cuyo título es: de los tres elementos, aire, agua y tierra, en que se trata de las cosas que en cada uno de ellos, acerca de las Occidentales Indias, engendra y produce comunes con las de aca y particulares de aquel nuevo mundo: Va dividido en tres partes, dando a cada uno de estos elementos, por el orden que aqui se exponen, su particular tratado y parte," in *Tomás López Medel: Trayectoria de un clérigo-oidor ante el nuevo mundo*, ed. Berta Ares-Queija. For the dates of this treatise, see Ares-Queija, *Tomás López Medel*, pp. 204–207.

45. The 1571 statutes were the product of Juan de Ovando's *visita* to the council in 1569. The Statutes for the Council of Indies were promulgated on September 24, 1571, in El Pardo. See Schäfer, *El Consejo Real*, vol. 1, pp. 130–135; Goodman, *Power and Penury*, p. 68ff.; Alvarez Peláez, *La conquista*, p. 131ff.; Jiménez de la Espada, *Relaciones*, vol. 1, p. 59; for the statutes of 1573, see Ordenanzas para la formación del libro de las

descripciones de Indias, July 3, 1573, AGI, Indiferente 427, L. 29, ff. 5v–66v; also in Solano, *Cuestionarios,* p. 16ff.

46. For more on Ovando's life, see Stafford Poole, "Juan de Ovando's Reform of the University of Alcalá de Henares, 1564–1566," *Sixteenth Century Journal* 21, no. 4 (1990): 585–586; A. W. Lovett, "Juan de Ovando and the Council of Finance (1573–1575)," *Historical Journal* 15, no. 1 (1972): 5ff.

47. Al licenciado Paulo de Laguna, Presidente del Real y Supremo Consejo de las Indias, Carta-prefacio de Antonio de Herrera, in Herrera, *Historia general,* vol. 1; Goodman, *Power and Penury,* p. 68ff.; Alvarez Peláez, *La conquista,* p. 131ff.; Jiménez de la Espada, *Relaciones,* vol. 1, p. 59. The Casa had its own system of information.

48. Schäfer, *El Consejo Real,* vol. 1, p. 129ff. See also Jiménez de la Espada, *Relaciones,* vol. 1, p. 59.

49. *CODOIN,* series 1, vol. 16, pp. 457–459.

50. Copy of Ordenanzas del Consejo de Indias, n.d., AGI, Indiferente 856, número 3: "porque ninguna cosa puede ser entendida ni tratada como debe cuyo sujeto no fuere primero sabido de las personas que de ella hubiere de conocer y determinar ordenamos y mandamos que los del nuestro consejo de las Indias con particular estudio y cuidado procuren tener hechas siempre descripciones y averiguaciones cumplida y cierta de todas las cosas del estado de las indias asi de la tierra como de la mar naturales y morales perpetuas y temporales eclesiasticas y seglares pasadas y presentes que por tiempo seran sobre que puede caer gobernacion / o disposicion de ley segun la orden e forma del titulo de las descripciones haciendolas ejecutar continuamente con mucha diligencia y cuidado." See also *Recopilación de leyes de los Reynos de las Indias,* vol. 2, libro 2, título 2, ley VI.

51. Real Cédula a los oficiales de la audiencia de Quito, August 16, 1572, El Escorial, in Solano, *Cuestionarios,* p. 15f.: "Sabed que deseando que la memoria de los hechos y cosas acaecidas en esas partes se conserva y que en el nuestro Consejo de las Indias haya noticia que debe haber de ellas y de las otras cosas de esas partes que son dignas de saberse, hemos proveído persona a cuyo cargo sea recopilarlos y hacer historia de ellas, por lo cual os encargamos que con diligencia os hagáis luego informar de cualesquier personas, así legas como religiosas, que en el distrito de esa audiencia hubieren escrito o recopilado, o tuvieren en su poder alguna historia, comentarios o relaciones de alguno de los descubrimientos, conquistas, entradas, guerras o facciones de paz y de guerras que en esas provincias o en partes de ellas hubiere habido desde su descubrimiento hasta los tiempos presentes. Y así mismo de la religión, gobierno, ritos y costumbres que los indios han tenido y tienen, y de la descripción de la tierra, naturaleza y calidades de las cosas de ella, haciendo así mismo buscar lo susodicho o algo de ello en los archivos, oficios y escritorios de los escribanos de gobernación y otras partes a donde pueda estar y lo que se hallare originalmente si se puediere. Y si no, la copia de ello daréis orden se nos envíe en la primera ocasión de flota o navíos que para estos reinos venga."

52. *Recopilación de leyes,* vol. 2, libro 4, título 1, ley VII (Statutes for New Discoveries and Settlements of 1573).

53. Ordenanzas para la formación del libro de las descripciones de Indias, July 3, 1573, AGI, Indiferente 427, L. 29, ff. 5v–66v; also in Solano, *Cuestionarios*, p. 16ff.

54. Ibid.

55. Real Cédula, 1573/12/10, El Pardo, AGI Indiferente 427, L. 29, ff. 95v–96r.

56. Jiménez de la Espada, *Relaciones*, vol. 1, p. 48f.

57. Jiménez de la Espada, one of the best-informed researchers on this subject, was unable to find any answers to this questionnaire. See ibid., p. 50. Spain was not the only state interested in surveying its territory. The ruler of Saxony, Elector Christian I (who ruled 1585–1591), ordered a survey of the Electorate in 1586.

58. Ibid., p. 59; see the "Instruction and questionnaire" of 1584 in Rene Acuña, "Instrucción y memoria de las relaciones que se han de hacer para la descripción de las Indias que su majestad manda hacer, para el buen gobierno y ennoblecimiento dellas," in *Relaciones geográficas del siglo XVI*, ed. Rene Acuña, vol. 1, pp. 73–78.

59. Instrucción y memoria de las relaciones para la descripción de las Indias, May 25, 1577, San Lorenzo del Escorial, AGI, Patronato 294, No. 11. This document is a manuscript copy contained in the actual report sent by the officials of Nueva Segovia, Venezuela. For a printed edition of the Instrucción, see the facsimile edition published by Carlos Sanz López, ed., *Relaciones geográficas de España y de Indias*.

60. Instrucción y memoria, AGI, Patronato 294, No. 11, f. 3r.

61. See *Recopilación de leyes*, vol. 2, libro 4, título 1, ley VII.

62. Thus these chapters preceded John Graunt's "Bills of Mortality" in London by almost a century. Although the Spanish accounts did not itemize particular individuals, they correlated numbers and causes. See John Graunt, *Reflections on the weeekly [sic] bills of mortality for the cities of London and Westminster, and the places adjacent . . . : but more especially, so far as they relate to the plague and other mortal diseases that we English-men are most subject unto: with an exact account of the greatest plagues that have happened since the creation*. Thanks to Paula Findlen for this reference.

63. Niculoso de Fornee, Pedro de Plascencia, Juan de Luque, and Francisco de Gallegos, "Descripción de la Tierra del Corregimiento de Abancay (1586)," in Jiménez de la Espada, *Relaciones*, vol. 2, p. 16: The *corregidor* Niculoso de Fornee "hizo parecer ante sí a Pedro de Plasencia y a Juan de Luque, españoles, y dijeron haber estado y residido y al presente residen en este valle de Xaxaguana; y el dicho Pedro de Plasencia dijo haber más de cuarenta años questá en este reino, y el dicho Juan de Luque haber más de veinte; y asímismo hizo parecer a Francisco de Gallegos, mestizo deste reino del Cuzco, el cual dijo haberse criado en este valle. De todos tres los susodichos recibí juramento en forma de derecho, y prometieron de decir verdad, y confirmándome yo el dicho corregidor con los pareceres de los susodichos acerca de los capítulos de la discrepción que Su Excelencia . . . [manda hacer de estos reinos del Cuzco]."

64. Luis de Monzón, Pedro Gonzáles, and Juan de Arbe, "Descripción de la Tierra del repartimiento de San Francisco de Atunrucana (1586)," in Jiménez de la Espada, *Relaciones*, vol. 1, p. 226: "yo Luis de Monzón, corregidor por Su Majestad desta provincia de Rucanas y Soras, habiendo recibido la Instrucción y Memoria en molde para hacer las relaciones de la descripción de las Indias que Su Majestad manda hacer para

el buen gobierno y ennoblecimiento dellas . . . ; y en su complimiento me junté en uno con Pedro Gonzáles, cura deste dicho pueblo, y con Juan de Arbe, español que ha residido mucho tiempo en esta provincia y repartimiento, y todos tres respondimos a los dichos capítulos en la forma siguiente; y para lo que se ha de saber de los indios, se hallaron presentes por lenguas Juan Alonso de Badajoz, mestizo, y el dicho Juan de Arbe, que la entiende, y ansímismo se hallaron presentes en esta junta los curacas y principales deste dicho repartimiento, que son don Juan Guancarilla y don Cristóbal Auchuqui, y don Francisco Hernández y don Francisco Curiaymara y otros caciques e indios principales."

65. Relación de Atlitlalaquia, February 22, 1580, Mexico City, in Acuña, *Relaciones,* vol. 6, t. 1, pp. 57 and 66; and Acuña's comment on this report, p. 55f.

66. Acuña, *Relaciones,* vol. 7, t. 2, p. 28.

67. Cline, "The Relaciones Geográficas," p. 193ff.

68. Nalle, *God in La Mancha,* p. 122.

69. *Recopilación de leyes,* vol. 2, libro 2, título 2, ley VI.

70. López Piñero, *Ciencia y técnica,* p. 167.

71. Frank Lestringant, *Mapping the Renaissance World: The Geographical Imagination in the Age of Discovery,* trans. David Fausett, p. 3. I am following Lestringant here.

72. Ibid., p. 4. On this so-called cosmographic revolution, see also Cirilo Flórez Miguel, "Ciencia y Renacimiento en la Universidad de la Salamanca," in Fernán Pérez de Oliva, *Cosmographia nova,* ed. Cirilo Flórez Miguel et al., p. 15ff.

73. Lestringant, *Mapping,* p. 45.

74. For more on the academy, see Mariano Esteban Piñeiro and María Isabel Vicente Maroto, "La Casa de la Contratación y la Academia Real Matemática," in *Historia de la ciencia,* ed. López Piñero. On the general European context of mapping and cosmographical projects, see Buisseret, *Monarchs, Ministers, and Maps.*

75. Observaciones de Juan López de Velasco sobre la jornada de Jaime Juan, January, 1583, AGI, Indiferente 740, No. 103: "El fin de la jornada para que se propone Jayme Juan es tomar las alturas o elevaciones de los lugares por do fuere."

76. Consulta del Consejo, February 5, 1583, Madrid, AGI, Indiferente 740, No. 103.

77. Instrucciones a Jaime Juan, n.d. [between February and April 1583?], AGI, Indiferente 740, No. 103.

78. Consulta del Consejo, July 28, 1595, Madrid, AGI, Indiferente 868.

79. In 1665 Robert Boyle drafted a natural history questionnaire for the use of travelers and navigators. See Robert Boyle, "General Heads for a Natural History of a Country, Great or Small," *Philosophical Transactions* 1 (1665–1666): 186–189; see also Margaret T. Hodgen, *Early Anthropology in the Sixteenth and Seventeenth Centuries,* pp. 188–190.

80. Alvarez Peláez, *La conquista,* p. 171.

81. See Charles B. Schmitt, "Experience and Experiment: A Comparison of Zabarella's View with Galileo's in *De Motu,*" *Studies in the Renaissance* 16 (1969): 80–138: "Zabarella, following Aristotle and the long Aristotelian tradition, saw it [nature] as a

living, biological entity, teleologically oriented and best understood through experience and syllogistic reasoning" (p. 124); Eugenio Garin, *Italian Humanism: Philosophy and Civic Life in the Renaissance*, p. 191ff.

5. BOOKS OF NATURE

1. Relato sobre lo que pasó con la flota de marzo de 1579, March 1579, AGI, Indiferente 1095, R. 24, No. 291.

2. Carta del obispo de Tucumán, March 12, 1579, Sanlúcar de Barrameda, AGI, Indiferente 1095, R. 24, No. 290: "necesariamente si dios no provee con un poco de tiempo más favorable están [las naos] en peligro muy grande y como jugadas a los dados. . . ."

3. Cristóbal Colón, "Carta a Santángel (1493)," in *Cristóbal Colón: Textos y documentos completos: Nuevas cartas,* ed. Consuelo Varela and Juan Gil, p. 221.

4. See Petrus Martyr de Angleria, "De Orbe Novo Decades Octo," in *Petrus Martyr de Angleria: Opera,* ed. Erich Woldan, Decade V.

5. See "Carta de Alonso de Zauzo al emperador Carlos V, January 22, 1518," in Jiménez de la Espada, *Relaciones,* vol. 1, p. 11ff.

6. See Hernán Cortés, *Cartas de relación de la conquista de Méjico.* There is an English translation of some of Cortés's letters: Anthony Pagden, ed., *Hernán Cortés: Letters from Mexico.*

7. See Gonzalo Fernández de Oviedo y Valdés, *Oviedo dela natural hystoria delas Indias;* also known as the *Sumario de la natural historia de las Indias* (hereafter cited as *Sumario*).

8. See Monardes, *Historia medicinal,* f. 1r/v.

9. For more on curiosities and wonders, see O. R. Impey and Arthur MacGregor, *The Origins of Museums: The Cabinet of Curiosities in Sixteenth- and Seventeenth-Century Europe;* Nicholas Jardine, James A. Secord, and E. C. Spary, *Cultures of Natural History;* Joy Kenseth, ed., *The Age of the Marvelous;* and Lorraine Daston and Katherine Park, *Wonders and the Order of Nature, 1150–1750.*

10. Angleria, "De Orbe Novo," Decade 1, Book 1, Chapter 9, f. 18r.

11. Acosta, *Historia natural,* pp. 129, 141.

12. Simón, *Noticias historiales,* vol. 1, p. 107: "Pero haber sido engaño este [de la zona tórrida] con otros muchos que tuvo Aristóteles, bien lo dice la experiencia de verlo habitado [el Nuevo Mundo]."

13. On this topic, see Grafton, *New Worlds, Ancient Texts;* Findlen, *Possessing Nature;* and Anthony Pagden, *European Encounters with the New World from the Renaissance to Romanticism.*

14. In 1535 Oviedo finished the first part of his *Historia;* it was published that year and in 1547; in 1557, the year of Oviedo's death, Book 20 of the first part was published. See Louise Bénet Tachot, "Comentarios acerca de la trayectoria editorial de *La Histo-*

ria General y Natural de las Indias de Gonzalo Fernández de Oviedo y Valdés," paper delivered at the 49th International Congress of Americanists.

15. Oviedo, *Historia,* vol. 1, p. 8.

16. Oviedo, *Sumario,* p. 4.

17. Alvarez Peláez, *La conquista,* p. 44.

18. Oviedo, *Historia,* vol. 1, p. 11.

19. Gonzalo Fernández de Oviedo, *Quincuagenas de los generosos e illustres e no menos famosos reyes, príncipes, duques, marqueses y condes, caballeros e personas notables de España,* Part 3, est. 23, quoted in Juan Pérez de Tudela, "Vida y escritos de Gonzalo Fernández de Oviedo," in Oviedo, *Historia,* vol. 1, p. xxxi.

20. See Pérez de Tudela, "Vida y escritos," pp. xx and xxi.

21. Ibid., pp. xvi–xxxiii.

22. For more on Oviedo's life, see ibid.; Gerbi, *Nature in the New World;* the collection of essays on Oviedo's life and work in José María de la Peña Cámara, ed., *Contribuciones documentales y críticas para una biografía de Gonzalo Fernández de Oviedo;* Enrique Otte, "Aspiraciones y actividades heterogéneas de Gonzalo Fernández de Oviedo, cronista," *Revista de Indias* 71 (1958): 9–62.

23. For more on the panther, see Oviedo, *Sumario,* p. 39ff.; on the pineapple, see Acosta, *Historia natural,* p. 260.

24. José de Sigüenza, *Historia de la Orden de San Jerónimo,* vol. 2, p. 151.

25. Paula Findlen, "Courting Nature," in *Cultures of Natural History,* ed. Jardine et al.

26. Oviedo, *Sumario,* p. 49; see also Pérez de Tudela, "Vida y escritos," p. lix.

27. Schäfer, *El Consejo Real,* vol. 1, p. 33f.

28. See Bartolomé de las Casas, *Brevísima relación de la destrucción de Indias.*

29. Consulta del Consejo, May 27, 1532, Medina del Campo, AGI, Indiferente 737, No. 24.

30. Ibid.

31. *CODOIN,* series 2, vol. 17, p. 288, August 18, 1532. This document is missing in the AGI.

32. Real Cédula al gobernador de la isla Fernandina, October 15, 1532, Segovia, AGI, Indiferente 422, L. 15, ff. 189v–189r.

33. Real Cédula al gobernador de la isla Fernandina, October 15, 1532, Segovia, AGI, Indiferente 422, L. 15, ff. 189v–189r.

34. Oviedo, *Sumario,* p. 4.

35. Edmundo O'Gorman argues that Oviedo discovered the nature of the New World because he was the first to write its natural history; see Edmundo O'Gorman, "Gonzalo Fernández de Oviedo y Valdés," in *Cuatro historiadores de Indias, siglo XVI: Pedro Mártir de Anglería, Gonzalo Fernández de Oviedo y Valdés, Bartolomé de las Casas, Joseph de Acosta,* p. 54. Before Oviedo's text, there were already other printed documents with reports on the novelty of the New World's nature, such as the letters of Columbus, Vespucci, and Cortés, the *Decades* of Peter Martyr, or the *Suma de geografía* of Martín Fernández de Enciso. See also Pérez de Tudela, "Vida y escritos," p. c.

36. Oviedo, *Sumario,* p. 84.

37. Ibid., p. 4.

38. Oviedo, *Historia,* vol. 1, p. 11.

39. Ibid., p. 9.

40. For more on the guaiacum and the hog plum, see Oviedo, *Sumario,* p. 75f.

41. Oviedo explained that his idea for this experiment came from Pliny and his reference to a type of mute frog that sings when brought to another land; Oviedo, *Historia,* vol. 2, p. 30.

42. Ibid., p. 46; for other experiments, see vol. 1, p. 174.

43. Oviedo, *Sumario,* p. 45.

44. Oviedo, *Historia,* vol. 1, p. 290.

45. Oviedo, *Sumario,* pp. 69 (coconuts), 66 (spiders).

46. Oviedo, *Historia,* vol. 1, p. 33f. He discussed the navigation to the Indies and concluded his remarks with these comments: "Esto que he dicho [sobre la navegación] no se puede aprender en Salamanca ni en Boloña, ni en París, sino en la cátedra de la gisola (que es aquel lugar donde va puesta el aguja de navegar), e con el cuadrante en la mano, tomando en la mar ordinariamente, las noches el estrella, e los días el sol con el astrolabio."

47. For information on the influence of Pliny in this period, see Charles G. Nauert, Jr., "Caius Plinius Secundus," in *Catalogus Translationum et Commentariorum: Medieval and Renaissance Latin Translations and Commentaries, Annotated Lists and Guides,* ed. Edward Cranz et al.; see Jean Céard, *La nature et les prodiges: L'Insolite au XVIe siècle,* p. 60ff.

48. Oviedo, *Historia,* vol. 1, p. 11; Oviedo, *Sumario,* p. 3.

49. Oviedo, *Historia,* vol. 1, p. 13.

50. Oviedo, *Sumario,* p. 37: "Verdad es que, según las maravillas del mundo y los extremos que las criaturas, más en unas partes que en otras, tiene, según las diversidades de las provincias y constelaciones donde se crían, ya vemos que las plantas que son nocivas en unas partes, son sanas y provechosas en otras, y las aves que en una provincia son de buen sabor, en otras ni las comen; los hombres, que en una parte son negros, en otras provincias son blanquísimos, y los unos y los otros son hombres. . . . Todas estas cosas, y otras muchas que se podrían decir a este propósito, son fáciles de probar y muy dignas de creer de todos aquellos que han leído o andado por el mundo, a quien la propia vista habrá enseñado la experiencia de lo que es dicho." See also *Historia,* vol. 2, pp. 39–40.

51. Oviedo, *Sumario,* p. 87.

52. Oviedo, *Historia,* vol. 2, p. 29; see also vol. 1, p. 277f.

53. Oviedo, *Historia,* vol. 1, p. 156; *Sumario,* p. 28.

54. Oviedo, *Historia,* vol. 1, p. 40.

55. Ibid., p. 41ff.

56. Ibid., pp. 41, 43.

57. Acosta, *Historia natural,* p. 64.

58. Ibid., p. 58.

59. Edmundo O'Gorman, "Segundo apéndice," in José de Acosta, *Historia natural y moral de las Indias,* ed. Edmundo O'Gorman, p. lxii.

60. For more on Acosta's life, see José Alcina Franch, "Introducción," in José de Acosta, *Historia natural y moral de las Indias;* and "Acosta, José de," in López Piñero et al., *Diccionario histórico,* vol. 1, p. 22f.

61. Alcina Franch, "Introducción," p. 10ff.

62. Letter from Father Gil González Dávila to Father Claudio Aquaviva, October 9, 1587, quoted in ibid., p. 16.

63. Acosta's *Historia* was a successful book: it was published again in Spanish (Barcelona, 1592; Seville, 1591; Madrid, 1610); Italian (Venice, 1596); French (Paris, 1598, 1605, 1606, 1616, 1661); German (Cologne, 1598, 1600, 1615; Ursel, 1605; Frankfurt, 1617); English (London, 1598, 1604, 1684); Dutch (Enkhnizen, 1598, 1624); and Latin (Frankfurt, 1590–1634). See Alcina, "Introducción," p. 41f.

64. Acosta, *Historia,* pp. 61–152.

65. Ibid., pp. 153–215 and 217–305.

66. Ibid., p. 57.

67. Ibid.

68. Ibid.

69. Ibid., p. 9.

70. Chaves, *Tratado de la sphera,* f. 10v.

71. See the Relación de Benito de Morales, n.d. [around 1570?], AGI, Indiferente 856; Real Cédula a Pedro Juan de Lastanosa, August 24, 1569, Madrid, AGI, Indiferente 426, L. 25, ff. 17r–18v.

72. See *Diccionario de autoridades* (Madrid: Gredos, 1963), "Machina: [third entry] Se llama también un todo compuesto artificiosamente de muchas partes heterogéneas, con cierta disposición que las mueve u ordena: por cuya semejanza se llama así el universo" (Machine: The name of a whole composed artificially of many different parts, with a certain disposition that moves and organizes them; by similarity the universe is also called a machine).

73. Acosta, *Historia,* p. 62.

74. Ibid., p. 241.

75. Ibid., p. 218.

76. Ibid., p. 139.

77. Ibid., p. 218.

78. Ibid., p. 178.

79. Ibid., p. 61.

80. Ibid., p. 61ff.

81. Ibid., p. 127.

82. Ibid., pp. 63, 58.

83. Ibid., p. 79.

84. Ibid. (passim).

85. Ibid., p. 78.

86. Ibid., p. 79.

87. Ibid.

88. Ibid., p. 78.

89. Ibid., p. 107.

90. Ibid., p. 68.

91. Ibid., p. 68f.

92. Ibid., p. 96ff.

93. Ibid., p. 98.

94. Real Cédula del Rey don Carlos, January 29, 1525, Madrid, AGI, Contratación 5787, No. 1, L. 1. ff. 33–34v.

95. Oviedo, *Historia,* vol. 2, p. 55.

96. Ibid., p. 35.

97. Ibid., p. 39.

98. See Medina, *El Veneciano,* vol. 1, p. 182.

99. Oviedo, *Historia,* vol. 1, p. 223.

100. Ibid., p. 235 (*batatas*); p. 236 (chiles); p. 230 (maize). For information on practical gardening, see Andrew Cunningham, "The Culture of Gardens," in *Cultures of Natural History,* ed. Jardine et al., pp. 53–55.

101. Carta del licenciado Maldonado al Rey, August 30, 1554, La Española, AGI, Contratación 5103.

102. Instrucciones al Dr. Francisco Hernández, November 1, 1570, Madrid, AGI, Indiferente 1228.

103. Carta del Dr. Hernández, September 1, 1574, México City, AGI, México 69, R. 6, No. 98.

104. Consulta del Consejo, November 20, 1578, Madrid, AGI, Chile 1, No. 4.

105. Daston and Park, *Wonders,* p. 136f.

106. For a parallel development in Europe, see Findlen, *Possessing Nature,* especially Chapter 6; Daston and Park, *Wonders,* pp. 137–141.

107. See Benavides, *Secretos de chirurgia,* f. 21r and f. On professionals collecting nature, see Findlen, *Possessing Nature,* pp. 4, 51, 70, 95, 201, 246, 271, and 286.

108. Jiménez Castellanos, "Prólogo," p. vii.

109. Ares-Queija, *Tomás López Medel,* p. 26. López Medel would use later the work of Monardes.

110. Jiménez Castellanos, "Prólogo," pp. vii–viii.

111. For more on Monardes, see Francisco Guerra, *Nicolás Bautista Monardes, su vida y su obra;* López Piñero, "Las 'nuevas medicinas' americanas"; Jiménez Castellanos, "Prólogo"; Eli de Gortari claims that Nahuatl medicinal knowledge reached Europe through Monardes; see Eli de Gortari, *La ciencia en la historia de México,* p. 172f.

112. Monardes, *Historia,* f. 1v.

113. López Piñero and Pardo Tomás, "The Contribution of Hernández," p. 127.

114. See Goodman, *Power and Penury,* p. 238. In the sixteenth century, buying and selling collections of wonders was a rare practice; it became more common during the

seventeenth century. See Paula Findlen, "Inventing Nature: Commerce, Art, and Science in the Early Modern Cabinet of Curiosities," in *Merchants and Marvels,* ed. Smith and Findlen, p. 300f.

115. Ignacio Jordán de Asso, *CL. Hispaniiensium atque exterorum epistolae cum praefatione et notis Ignatii de Asso,* p. 53.

116. For a description of the Argote museum, see Monardes, *Historia medicinal,* p. 81; and Francisco Pacheco, *Libro de descripción de verdaderos retratos, de illustres y memorables varones por . . . ,* p. 72ff.

117. Monardes, *Historia medicinal,* p. 81.

118. Pacheco, *Libro de descripción,* p. 72ff.; on the different functions of Renaissance gardens, see Cunningham, "The Culture of Gardens," pp. 38–56.

119. Monardes, *Historia medicinal,* p. 135.

120. Ibid., p. 79.

121. *Colección Navarrete,* vol. 16, p. 279.

122. This information comes from Chaves's postmortem inventory elaborated by Professor Klaus Wagner based on documentation from the Archivo de Protocolos de Sevilla. I am very grateful to Professor Wagner for letting me have a copy of this part of the inventory. "Inventario post-mortem de Jerónimo de Chaves" (I only translated part of it; this is the complete list):

un pedaço de vallena,
tres palos traydos de las Yndias para el yjada,
quatro borcelanas [*sic*] a manera de escudillas de la Yndia de Portugal,
un gueuo de avestruz,
nueue aras de piedra quajada traidas de Nueva España de diversos colores,
un concha de tortuga y un caracol,
una caja llena de Mechoacan,
una caja llena de conchas,
tres arcos turquescos con un aljaba llena de saetas,
otros dos maças de . . . turquescas,
tres arcos yngleses y de yndios,
una fuenta grande de calabaça de Yndias,
cinco borbones de Yndias,
un bote de uidrio con arcachofas de las yndias en conserba,
una redoma con balçamo de Yndias.

123. Carta de Juan de Castañeda a Carolus Clusius, October 20, 1600, Sevilla, in Asso, *CL. Hispaniiensium,* p. 56.

124. For more on the Spanish soldier Pedro de Osma, see below and note 127. For more on the English explorer, see Hakluyt, *The Principal Navigations,* vol. 6, p. 170.

125. Carolus Clusius, *Rariorum plantarum historia,* libro 2, p. 173.

126. Daston and Park, *Wonders,* p. 158.

127. Carta de Pedro de Osma al Dr. Monardes, December 26, 1568, Lima, in Monardes, *Historia medicinal,* p. 73ff.

128. Carolus Clusius, *Exoticorum libri decem,* pp. 351–353.

129. Carta de Juan de Castañeda to Carolus Clusius, in Asso, *CL. Hispaniiensium,* p. 53.

130. The letters of Castañeda to Clusius are in ibid., pp. 53–70.

131. For more on Clusius's interest in *costo,* see Findlen, *Possessing Nature,* p. 270.

132. Asso, *CL. Hispaniiensium,* p. 70.

CONCLUSIONS: THE POLITICS OF KNOWLEDGE

1. Fernández de Enciso, *Suma de geografía,* p. 202.

2. Pedro de Medina, *Arte de navegar en que se contienen todas las reglas, declaraciones, secretos y avisos, que a la buena navegación son necessarios, y se deven saber.*

3. Pedro de Medina, *Regimiento de Navegación. En que se contienen todas las reglas, declaraciones y avisos del libro del arte de navegar;* López Piñero et al., *Diccionario histórico,* vol. 2, p. 49.

4. Cortés, *Breve compendio de la sphera.*

5. See López Piñero, *Ciencia y técnica,* pp. 202ff. and 205f.

6. López Piñero et al., *Diccionario histórico,* vol. 1, p. 218f.; Karrow, *Mapmakers,* pp. 116–117.

7. López Piñero, *Ciencia y técnica,* p. 203.

8. E. G. R. Taylor, ed., *A Regimento for the Sea and Other Writings on Navigation by William Bourne of Gravesend, a Gunner (c. 1535–1582),* p. 6f.

9. Hakluyt, *Voyages,* vol. 1, p. 16f.

10. Jean Marc Pelorson, *Les "Letrados": Juristes castillans sous Philippe III: Recherches sur leur place dans la société, la culture, et l'état;* A. Conavas del Castillo, *Estudios del reinado de Felipe IV,* 2 vols. (Madrid: A. Pérez Dubrull, 1888); John Elliott, *The Count-Duke of Olivares: The Statesman in an Age of Decline,* and his *Imperial Spain, 1469–1716;* David Vassberg, *Land and Society in Golden Age Castile;* Henry Kamen, *Spain in the Later Seventeenth Century, 1665–1700;* R. A. Stradling, "Seventeenth Century Spain: Decline or Survival?" *European Studies Review* 9 (1979): 157–194; Henry Kamen, "The Decline of Spain: A Historical Myth?" *Past and Present* 81 (November 1978): 24–50; J. I. Israel, "Debate—The Decline of Spain: A Historical Myth?" *Past and Present* 91 (May 1981): 170–180; Thompson, *War and Society in Habsburg Spain;* Maravall, *Estado moderno.*

11. Cárdenas, *Problemas y secretos;* Alvaro Alonso Barba, *Arte de los metales en que se enseña el verdadero beneficio de los de oro y plata por açogue. El modo de fundirlos todos, y como se han de refinar y apartar unos de otros.*

Bibliography

ARCHIVAL SOURCES

Berkeley, Bancroft Library
Proceso de la Inquisición contra Miguel Redelic, alias Miguel Alemán
Madrid, Archivo Histórico Nacional
Documentos Diversos 22
Seville, Archivo General de Indias (AGI)
Chile 1
Contaduría 672
Contratación 5784, 5787
Escribanía 1008c
Indiferente 418, 419, 420, 421, 422, 423, 426, 427, 737, 738, 740, 746, 856, 857, 868, 1092, 1095, 1204, 1223, 1228, 1528, 1961, 1962
Lima 28b, 569
México 19, 69, 109, 168, 2999
Panamá 233
Pasajeros 2
Patronato 20, 34, 174, 238, 251, 259, 264, 294
Santo Domingo 99
Seville, Biblioteca Colombina
Manuscritos
Washington, Library of Congress
Manuscripts, Kraus Collection

PRINTED PRIMARY SOURCES

[At the Biblioteca Colombina, the Biblioteca Nacional de Madrid, the Biblioteca de la Universidad de Sevilla, the Bancroft Library, the Biblioteca de Medicina de la Universidad Complutense (Madrid), the Special Collection at Stanford University Library, and the John Carter Brown Library (Providence) I found some of the printed sources mentioned below.]

Colección de documentos inéditos relativos al descubrimiento, conquista y colonización de las posesiones españolas en América y Oceanía (CODOIN). 42 vols. Nendeln, Liechtenstein: Kraus Reprint, 1964–1969.

Colección de documentos y manuscriptos compilados por Fernández de Navarrete. 32 vols. Nendeln, Liechtenstein: Kraus-Thomson Organization Limited, 1971.

Recopilación de leyes de los Reynos de las Indias. 4 vols. Madrid: Julián de Paredes, 1681.

Acosta, José de. *Historia natural y moral de las Indias.* Ed. Edmundo O'Gorman. Mexico City: Fondo de Cultura Económica, 1962.

———. *Historia natural y moral de las Indias.* Madrid: Historia 16, 1987.

Acuña, Rene, ed. *Relaciones geográficas del siglo XVI.* 10 vols. Mexico City: UNAM, 1982.

Angleria, Petrus Martyr de. "De Orbe Noveo Decades Octo." In *Petrus Martyr de Angleria: Opera,* ed. Erich Woldan. Graz: Akademische Druck- u. Verlagsanstalt, 1966. [Facsimile of the 1530 edition.]

Asso, Ignacio Jordán de. *CL. Hispaniiensium atque exterorum epistolae cum praefatione et notis Ignatii de Asso.* Caesaraugustae: Ex Typographia Regia, 1793.

Barba, Alvaro Alonso. *Arte de los metales en que se enseña el verdadero beneficio de los de oro y plata por açogue. El modo de fundirlos todos, y como se han de refinar y apartar unos de otros.* Madrid: Imprenta del Reino, 1640.

Benavides, Pedrarias de. *Secretos de chirurgia, especial de las enfermedades de Morbo galico y Lamparones y Mirrarchia, y asimismo la manera como se curan los Indios de llagas y heridas y otras passiones en las Indias, muy util y provechoso para en España y otros muchos secretos de chirurgia hasta agora no escritos.* Valladolid: Francisco Fernández de Cordona, impresor de la Magestad Real, 1567.

Boyle, Robert. "General Heads for a Natural History of a Country, Great or Small." *Philosophical Transactions* 1 (1665–1666): 186–189.

Cárdenas, Juan de. *Problemas y secretos maravillosos de las Indias.* Madrid: Ediciones Cultura Hispánica, 1945. [Facsimile of the first edition by Pedro Ocharte in Mexico City 1591.]

Casas, Bartolomé de las. *Brevísima relación de la destrucción de Indias.* Madrid: Fundación Universitaria Española, 1977. [Facsimile of the 1552 edition.]

Chaves, Jerónimo de. *Tractado de la sphera que compuso el doctor Ioannes de Sacrobusto con muchas adiciones. Agora nuevamente traduzido de Latin en lengua Castellana por el bachiller Hierónymo de Chaves: el qual añidio muchas figuras, tablas, y claras *demostraciones: juntamente con unos breves scholios, necessarios a mayor illucidation, ornato y perfection del dicho tratado.* Seville: Casa de Juan de León, 1545.

———. *Chronographia o Repertorio de los tiempos, el mas copioso y preciso que hasta agora ha salido a luz; en el qual se tocan y declaran materias muy provechosas de philosophia, astrologia, cosmographia y medicina. . . .* Seville, 1548.

Clusius, Carolus. *Rariorum plantarum historia.* Antwerp: Ex Officina Plantiniana Apud Ioannem Moretum, 1601.

———. *Exoticorum libri decem.* Antwerp?: Ex Officina Plantiniana Raphelengii, 1605.

Cortés, Hernán. *Cartas de relación de la conquista de Méjico.* 2 vols. Madrid: Espasa Calpe, 1940.

Cortés, Martín. *Breve compendio de la sphera y de la arte de navegar, con nuevos instrumentos y reglas, ejemplificado con muy subtiles demostraciones.* Seville: En casa de Anton Alvarez, 1551.

Eden, Richard. *The Decades of the newe worlde or west India, contynyng the navigations and conquestes of the Spanyardes, with the particular description of the moste ryche and large landes and Ilandes lately founde in the west Ocean perteynyng to the inheritaunce of the kinges of Spayne. In the which the diligente reader may not only consyder what commoditie may heraby chaunce to the hole chirstian world in tyme to come, but also learne many secreates touchynge the lande, the sea, and the starres, very necessarie to be knowen to al such as shal attempte any navigationes, or otherwise have delite to beholde the strange and wooderfull woorkes of God and nature. Written in the Latin tonge by Peter Martyr of Angleria, and translated into Englishe by Rycharde Eden.* London: Guilhelmi Powell, 1555.

Escalante de Mendoza, Juan de. *Itinerario de navegación de los mares y tierras occidentales.* Madrid: Museo Naval, 1985.

Farfán, Agustín. *Tractado breve de medicina.* Madrid: Ediciones Cultura Hispánica, 1944. [Facsimile of the edition printed in Mexico by Pedro de Ocharte in 1592.]

Fernández de Enciso, Martín. *Suma de geografía que trata de todas las partidas y provincias del mundo: en especial de las indias y trata largamente del arte del marear juntamente con la espera en romance: con el regimiento del sol y del norte: agora nuevamente emendada de algunos defectos que tenia la impression passada.* Seville: En casa de Juan Cromberg, 1530.

Fernández de Oviedo y Valdés, Gonzalo. *Oviedo dela natural hystoria delas Indias.* Toledo: Remon de Petras, 1526.

———. *Historia general y natural de las Indias.* Edited and introductory essay by Juan Pérez de Tudela Bueso. 5 vols. Madrid: Biblioteca de Autores Españoles, 1959.

Gesner, Conrad. *Evonymus C. Gesneri Medici de Remedis secretis, Liber Physicus, Medicus & partim etiam Chymicus, & Oeconomicus in vinorum diversi saporis apparatu, Medicis & Pharmacopolis omnibus praecipue necessarius, nunc primum in lucem editus.* [Zurich: Chr. Froschauer, ca. 1565] (place and date of publication according to the catalog of Klaus Wagner).

Graunt, John. *Reflections on the weeekly [sic] bills of mortality for the cities of London and Westminster, and the places adjacent . . . but more especially, so far as they relate to the plague and other mortal diseases that we English-men are most subject unto: with an exact account of the greatest plagues that have happened since the creation.* London: Printed for Samuel Speed, 1665.

Hakluyt, Richard. *The Principal Navigations, Voyages, Traffiques & Discoveries of the English Nation.* 12 vols. Glasgow: James McLehose and Sons, 1903.

Heredia Herrera, Antonio. *Catálogo de las Consultas del Consejo de Indias.* Vols. 1 and 2 [1529–1599]. Madrid: Dirección General de Archivos y Bibliotecas, 1972.

———. *Catálogo de las Consultas del Consejo de Indias.* Vols. 1 to 3 [1600–1616]. 9 vols. Seville: Diputación Provincial de Sevilla, 1983–84.

Herrera y Tordesillas, Antonio de. *The General History of the Vast Continent and Islands of America, Commonly call'd, The West-Indies, From the First Discovery thereof: With the Best Accounts the People could give of their Antiquities. Collected from the Original Relations sent to the Kings of Spain. By Antonio de Herrera, Historiographer of His Catholic Majesty. Translated into English by Capt. John Stevens.* 2 vols. London: Jer. Batley, 1725.

———. *Historia general de los hechos de los Castellanos en las Islas i Tierra Firme del Mar Océano escrita por Antonio de Herrera coronista mayor de su Magestad de las Indias y su coronista de Castilla, en quatro decadas desde el año de 1492 hasta el de 1531.* Madrid: n.p., 1730 [originally published 1601–1615].

Jane, Cecil. *The Four Voyages of Columbus.* 2 vols. New York: Dover Publications, Inc., 1988.

Jiménez de la Espada, Marcos. "Antecedentes." In *Relaciones geográficas de Indias,* vol. 2, pp. lxxvii–lxxxvii. 3 vols. Madrid: Biblioteca de Autores Españoles, 1965.

———. *Relaciones geográficas de Indias: Perú.* Vol. 183 of the collection. 3 vols. Madrid: Biblioteca de Autores Españoles, 1965.

Laguna, Andrés. *Pedacio Dioscórides Anazarbeo, acerca de la materia medicinal, y de los venenos mortíferos: Traduzido de la lengua Griega en la vulgar Castellana, e illustrado con claras y substanciales annotaciones, y con las figuras de unnúmeras plantas exquisitas y raras por el doctor . . . , médico de Julio III, Pontífece Máximo.* Anvers: En casa de Juan Latio, 1555.

López de Gómara, Francisco. *La istoria de las Indias, y conquista de Mexico.* Zaragoza: En casa de Agustin Millan, 1552.

López Medel, Tomás. "Tratado cuyo título es: de los tres elementos, aire, agua y tierra, en que se trata de las cosas que en cada uno de ellos, acerca de las Occidentales Indias, engendra y produce comunes con las de aca y particulares de aquel nuevo mundo: Va dividido en tres partes, dando a cada uno de estos elementos, por el orden que aqui se exponen, su particular tratado y parte." In *Tomás López Medel: Trayectoria de un clérigo-oidor ante el nuevo mundo,* ed. Berta Ares-Queija, 409–558. Guadalajara: Instituto Provincial de Cultura "Marqués de Santillana," 1993.

Medina, Pedro de. *Arte de navegar en que se contienen todas las reglas, declaraciones, secretos y avisos, que a la buena navegación son necessarios, y se deven saber.* Valladolid, 1545. Facsimile ed. Madrid: Talleres Helios, 1945.

———. *Regimiento de Navegacion. En que se contienen las reglas, declaraciones y avisos del libro del arte de navegar.* Seville: Juan Canalla, 1552.

Monardes, Nicolás Bautista. *[Primera y Segunda y Tercera partes de la] Historia Medicinal de las cosas que se traen de nuestras Indias Occidentales que sirven en Medicina.* Seville: Padilla Libros, 1988. [Facsimile of the 1574 edition.]

Morales, Garciperez. *Tratado del Balsamo y de sus utilidades para las enfermedades del cuerpo humano: Compuesto por el Doctor . . . cathedratico de prima en el colegio de Sancta Maria de Jesu de la ciudad de Sevilla. Dirigido al yllustrissimo señor don Pedro Giron Duque y Conde de Ureña.* Seville: En casa de Juan Varela, 1530.

Navagero, Andrea. *Viaje a España del magnífico señor Andrés Navagero (1524–1526) embajador de la República de Venecia ante el Emperador Carlos V.* Valencia: Editorial Castalia, 1951.

Pacheco, Francisco. *Libro de descripción de verdaderos retratos, de illustres y memorables varones. . . .* Seville, 1599. [Facsimile edition: n.p., n.d.]

Pagden, Anthony, ed. *Hernán Cortés: Letters from Mexico.* New Haven and London: Yale University Press, 1986.

Simón, Pedro. *Noticias historiales de las conquistas de Tierra Firme en las Indias Occidentales.* Vol. 1 [103 of the collection]. 6 vols. Bogotá: Biblioteca Banco Popular, 1981.

Tafur, Pero. *Andanças e viajes de Pero Tafur por diversas partes del mundo avidos (1435–1439).* Madrid: Imprenta de Miguel Ginesta, 1874.

Taylor, E. G. R., ed. *A Regimento for the Sea and Other Writings on Navigation by William Bourne of Gravesend, a Gunner (c. 1535–1582).* Cambridge: Cambridge University Press, 1963.

Varela, Consuelo, and Juan Gil, eds. *Cristóbal Colón: Textos y documentos completos—Nuevas cartas.* Madrid: Alianza Editorial, 1992.

Veitía Linage, Joseph de. *Norte de la Contratación de las Indias Occidentales.* Buenos Aires: Publicaciones de la Comisión Argentina de Fomento Interamericano, 1945. [1st ed.: Seville, 1672.]

SECONDARY SOURCES

Dictionnaire de l'Académie française. 1694. Available: http://duras.uchicago.edu/cgibin/quick_look.new.sh?word=tartane.

Albuquerque, Luis de. *Historia de la navegación portuguesa.* Madrid: Mapfre, 1991.

Albuquerque, Luís de, and Francisco Contente Domingues. *Dicionário de história dos descobrimentos portugueses.* Lisbon: Caminho, 1994.

Alvarez Peláez, Raquel. "La obra de Hernández y su recuperación ilustrada." In *La real expedición botánica a Nueva España, 1787–1803,* 147–158. Madrid: Real Jardín Botánico/Consejo Superior de Investigaciones Científicas, 1987.

———. *La conquista de la naturaleza americana.* Madrid: Consejo Superior de Investigaciones Científicas, 1993.

Arenas Frutos, Isabel. "Inventos sobre tecnología submarina para la América colonial." *Asociación de Historiadores Latinoamericanistas Europeos (AHILA),* special issue (1992): 421–434.

Ares-Queija, Berta. *Tomás López Medel: Trayectoria de un clérigo-oidor ante el nuevo mundo.* Guadalajara: Institución Provincial de Cultura "Marqués de Santillana," 1993.

Arranz Márquez, Luis. *Repartimiento y encomiendas en la Isla Española (El Repartimiento de Albuquerque de 1514).* Santo Domingo: Ediciones Fundación García Arévalo, 1991.

Bakewell, Peter. "Technological Change in Potosí: The Silver Boom of the 1570's." *Jahrbuch für Geschichte von Staat, Wirtschaft und Gesellschaft Lateinamerikas* 14 (1977): 57–77.

———. "Mining." In *Colonial Spanish America,* ed. Leslie Bethell, 203–249. Cambridge: Cambridge University Press, 1987.

Barber, Peter. "England II: Monarchs, Ministers, and Maps, 1550–1625." In *Monarchs, Ministers, and Maps: The Emergence of Cartography as a Tool of Government in Early Modern Europe,* ed. David Buisseret. Chicago and London: University of Chicago Press, 1992.

Bargalló, Modesto. *La minería y la metalurgia en la América española durante la época colonial.* Mexico City and Buenos Aires: Fondo de Cultura Económica, 1955.

Barrera, Antonio. "Local Herbs, Global Medicines: Commerce, Knowledge, and Commodities in Spanish America." In *Merchants and Marvels: Commerce, Science, and Art in Early Modern Europe,* ed. Pamela H. Smith and Paula Findlen. New York: Routledge, 2002.

Berthe, Jean-Pierre. "El cultivo del 'pastel' en Nueva España." *Historia Mexicana* 9, no. 3 (1960): 340–367.

Biringucci, Vannoccio. *The Pirotechnia of Vannoccio Biringucci: The Classic Sixteenth-Century Treatise on Metals and Metallurgy.* Trans. Cyril Stanley Smith and Martha Teach Gnudi. New York: Dover Publications, 1990.

Bitterli, Urs. *Cultures in Conflict.* Stanford: Stanford University Press, 1989.

Bouwsma, William J. "Anxiety and the Formation of Early Modern Culture." In *After the Reformation: Essays in Honor of J. H. Hexter,* ed. Barbara C. Malament, 215–246. Philadelphia: University of Pennsylvania Press, 1980.

Brading, David. *The First America: The Spanish Monarchy, Creole Patriots, and the Liberal State, 1492–1867.* Cambridge: Cambridge University Press, 1991.

———. *Orbe Indiano: De la Monarquía católica a la República criolla, 1492–1867.* Mexico City: Fondo de Cultura Económica, 1998.

Braudel, Fernand. *The Mediterranean and the Mediterranean World in the Age of Philip II.* 2 vols. New York: Harper, 1966.

Briesemeister, Dietrich. "La imagen de América en la Alemania que conoció Hernando Colón." In *Hernando Colón y su época,* 27–46. Seville: Real Academia Sevillana de Buenas Letras, 1991.

Brockway, Lucile H. *Science and Colonial Expansion: The Role of the British Royal Botanic Gardens.* New Haven, Conn., and London: Yale University Press, 2002.

Buisseret, David, ed. *Monarchs, Ministers, and Maps: The Emergence of Cartography as a Tool of Government in Early Modern Europe.* Chicago: University of Chicago Press, 1992.

Burke, Edmund. "Contested Paradigms in Early Modern World History." Paper presented at the University of California at Davis conference on Modernity's Histories in Global Context: Contested Narratives, Models, Processes, May 17–18, 1997.

Calero, L. F. *Chiefdoms under Siege: Spain's Rule and Native Adaptation in the Southern Colombian Andes, 1535–1700.* Albuquerque: University of New Mexico Press, 1997.

Carande, Ramón. *Carlos V y sus banqueros.* 2 vols. Barcelona: Editorial Crítica, 1977.

Castañeda Delgado, Paulino, Mariano Cuesta Domingo, and Pilar Hernández Aparicio. "Estudio preliminar." In *Alonso de Chaves: Quatri partitu en cosmografía práctica, y por otro nombre Espejo de navegantes.* Madrid: Instituto de Historia y Cultura Naval, 1983.

Castillo Martos, Manuel, and Mervyn Francis Lang. *Metales preciosos: Unión de dos mundos.* Seville and Bogotá: Muñoz Moya y Montraveta, 1995.

Castillo Quartiellerz, Rodolfo del. *Documento inédito del siglo XVI referente a D. Fernando Colón.* Madrid: Real Academia de la Historia, 1898.

Céard, Jean. *La nature et les prodiges: L'Insolite au XVIe siècle.* Geneva: Librairie Droz, 1996.

Cervera Vera, Luis. *Estudios sobre Juan de Herrera.* Madrid: Castalia, 1972.

Chaunu, Huguette, and Pierre Chaunu. *Séville et l'Atlantique (1504–1650).* 8 vols. Paris: Librairie Armand Colin, 1955.

Chevalier, François. *Land and Society in Colonial Mexico.* Berkeley and Los Angeles: University of California Press, 1972.

Cipolla, Carlo M. *Guns, Sails, and Empires: Technological Innovation and the Early Phases of European Expansion, 1400–1700.* New York: Pantheon Books, 1965.

Cline, Howard F. "The Relaciones Geográficas of the Spanish Indies, 1577–1648." In *Handbook of Middle American Indians,* ed. Howard F. Cline, vol. 12. Austin: University of Texas Press, 1972.

Cobo, Bernabé. *Obras.* 2 vols. Madrid: Atlas, 1964.

Cook, David N. *Born to Die: Disease and New World Conquest, 1492–1650.* Cambridge and New York: Cambridge University Press, 1998.

Cortesão, Armando. *Cartografia e cartógrafos portugueses dos séculos XV e XVI.* 2 vols. Lisbon: Seara Nova, 1935.

Crosby, Alfred. *The Columbian Exchange: Biological and Cultural Consequences of 1492.* Westport, Conn.: Greenwood Press, 1972.

———. *Ecological Imperialism: The Biological Expansion of Europe, 900–1900.* Cambridge and New York: Cambridge University Press, 1986.

Crosland, Maurice. "The Development of a Professional Career in Science in France." In *The Emergence of Science in Western Europe,* ed. Maurice Crosland, 127–159. London and Basingstoke: Macmillan Press Ltd., 1975.

Cuesta, Mariano. "El tratado de Tordesillas y la cartografía en la época de los reyes católicos." In *El tratado de Tordesillas en la cartografía histórica,* ed. Jesús Varela-Marcos, 53–84. Valladolid: Junta de Castilla y León, 1994.

Cuevas, Mariano. *Monje y marino: La vida de fray Andrés de Urdaneta.* Mexico City: Editorial Layac, 1943.

Cunningham, Andrew. "The Culture of Gardens." In *Cultures of Natural History,* ed. Nicholas Jardine, James A. Secord, and E. C. Spary. Cambridge and New York: Cambridge University Press, 1996.

Daston, Lorraine, and Katharine Park. *Wonders and the Order of Nature, 1150–1750.* New York: Zone Books, 1998.

Dear, Peter. *Revolutionizing the Sciences: European Knowledge and Its Ambitions, 1500–1700.* Princeton: Princeton University Press, 2001.

Debus, Allen G. *The Chemical Philosophy: Paracelsian Science and Medicine in the Sixteenth and Seventeenth Centuries.* Vol. 1. 2 vols. New York: Science History Publications (Neale Watson Academic Publications), 1977.

Díaz, José Simón. *Bibliografía de la literatura hispánica.* 16 vols. Madrid: Consejo Superior de Investigaciones Científicas, 1960–1994.

Domínguez Ortiz, Antonio. *The Golden Age of Spain, 1516–1659.* The History of Spain. New York: Basic Books, 1971.

———. "Instituciones políticas y grupos sociales en Castilla durante el siglo XVII." In *Instituciones y sociedad en la España de los Austrias.* Barcelona: Editorial Ariel, S.A., 1985.

Elliott, John H. "The Decline of Spain." *Past and Present* 20 (1961): 52–75.

———. *The Old World and the New.* Cambridge: Cambridge University Press, 1970.

———. *The Count-Duke of Olivares: The Statesman in an Age of Decline.* New Haven: Yale University Press, 1986.

———. *Imperial Spain, 1469–1716.* New York: Penguin, 1990.

Esteban Piñeiro, Mariano, and María Isabel Vicente Maroto. "La Casa de la Contratación y la Academia Real Matemática." In *Historia de la ciencia y de la técnica en la Corona de Castilla,* ed. José María López-Piñero, vol. 3, 35–51. 4 vols. Salamanca: Junta de Castilla y León Consejería de Educación y Cultura, 2002.

Fernández Alvarez, Manuel. *La sociedad española del Renacimiento.* Madrid: Cátedra, 1974.

———. *La sociedad española en el Siglo de Oro.* 2nd ed., rev. and enlarged. Madrid: Editorial Gredos, 1989.

———. *Felipe II y su tiempo.* Madrid: Espasa, 2002.

Findlen, Paula. *Possessing Nature: Museums, Collecting, and Scientific Culture in Early Modern Italy.* Berkeley: University of California Press, 1994.

———. "Courting Nature." In *Cultures of Natural History,* ed. Nicholas Jardine, James A. Secord, and E. C. Spary. Cambridge and New York: Cambridge University Press, 1996.

———. "Inventing Nature: Commerce, Art, and Science in the Early Modern Cabinet of Curiosities." In *Merchants and Marvels: Commerce, Science, and Art in Early Modern Europe,* ed. Pamela H. Smith and Paula Findlen. New York: Routledge, 2002.

Flórez Miguel, Cirilo. "Ciencia y Renacimiento en la Universidad de la Salamanca." In Fernán Pérez de Oliva, *Cosmographia nova,* ed Cirilo Flórez et al. Salamanca: Universidad de Salamanca, 1985.

Franch, José Alcina. "Introducción." In *José de Acosta: Historia natural y moral de las Indias.* Madrid: Historia 16, 1986.

García Tapia, Nicolás. *Técnica y poder en Castilla durante los siglos XVI y XVII.* Salamanca: Junta de Castilla y León, 1989.

Garin, Eugenio. *Italian Humanism: Philosophy and Civic Life in the Renaissance.* New York: Harper and Row, 1965.

Gerbi, Antonello. *Nature in the New World.* Pittsburgh: University of Pittsburgh Press, 1985.

Gimpel, Jean. *The Medieval Machine.* London: Penguin, 1977.

González, Francisco José. *Astronomía y navegación en España: Siglos XVI–XVIII.* Madrid: Mapfre, 1992.

Goodman, David C. *Power and Penury: Government, Technology, and Science in Philip II's Spain.* Cambridge: Cambridge University Press, 1988.

Gortari, Eli de. *La ciencia en la historia de México.* Mexico City, Barcelona, and Buenos Aires: Editorial Grijalbo, S.A., 1979.

Grafton, Anthony. *New Worlds, Ancient Texts: The Power of Tradition and the Shock of Discovery.* Cambridge, Mass.: Belknap Press of Harvard University Press, 1995.

Greenblatt, Stephen. *Marvelous Possessions.* Chicago: University of Chicago Press, 1991.

Greenleaf, Richard E. *Inquisición y sociedad en el México colonial.* Madrid: Ediciones José Porrúa Turanzas, S.A., 1985.

Grove, Richard H. *Green Imperialism.* Cambridge: Cambridge University Press, 1995.

Guerra, Francisco. *Nicolás Bautista Monardes, su vida y su obra.* Mexico City: Compañía Fundidora de Fierro y Acero de Monterrey, 1961.

Hanke, Lewis. *The Spanish Struggle for Justice in the Conquest of America.* Boston: Little, Brown and Company, 1965.

Haring, Clarence Henry. *Trade and Navigation between Spain and the Indies.* Cambridge, Mass.: Harvard University Press, 1918.

———. *El comercio y la navegación entre España y las Indias en época de los Habsburgos.* Paris and Brussels: Desclée, De Brouwer, 1939.

Hodgen, Margaret T. *Early Anthropology in the Sixteenth and Seventeenth Centuries.* Philadelphia: University of Pennsylvania Press, 1964.

Icaza, Francisco A. de. *Diccionario autobiográfico de conquistadores y pobladores de Nueva España.* 2 vols. Guadalajara: Edmundo Aviña Levy, 1969.

Impey, O. R., and Arthur MacGregor. *The Origins of Museums: The Cabinet of Curiosities in Sixteenth- and Seventeenth-Century Europe.* 2nd ed. London: House of Stratus, 2001.

Iñiguez Almech, Francisco. *Casas reales y jardines de Felipe II.* Madrid: Consejo Superior de Investigaciones Científicas, 1952.

Israel, J. I. "Debate—The Decline of Spain: A Historical Myth?" *Past and Present* 91 (May 1981): 170–180.

Jardine, Nicholas, James A. Secord, and E. C. Spary. *Cultures of Natural History.* Cambridge and New York: Cambridge University Press, 1996.

Jiménez Castellanos y Calvo Rubio, Juan. "Prólogo." In *Historia medicinal de las cosas que se traen de nuestras Indias Occidentales que sirven en medicina . . . por Nicolás Monardes,* v–xvii. Seville: Padilla Libros, 1988.

Kagan, Richard L. "Philip II and the Art of Cityscape." *Journal of Interdisciplinary History* 17, no. 1 (1986): 115–135.

———. "Clio and the Crown: Writing History in Habsburg Spain." In *Spain, Europe and the Atlantic World: Essays in Honour of John H. Elliott,* ed. Richard L. Kagan and Geoffrey Parker. Cambridge: Cambridge University Press, 1995.

Kamen, Henry. "The Decline of Spain: A Historical Myth?" *Past and Present* 81 (November 1978): 24–50.

———. *Spain in the Later Seventeenth Century, 1665–1700.* London and New York: Longman, 1980.

———. *Philip of Spain.* New Haven and London: Yale University Press, 1997.

Karrow, Robert W. *Mapmakers of the Sixteenth Century and Their Maps.* Chicago: Newberry Library, 1993.

Kenseth, Joy, ed. *The Age of the Marvelous.* Hanover, N.H.: Hood Museum of Art, 1991.

Lamb, Ursula S. "Science by Litigation: A Cosmographic Feud." *Terrae Incognitae* 1 (1969): 40–57.

———. "The Spanish Cosmographic Juntas of the Sixteenth Century." *Terrae Incognitae* 6 (1974): 51–64.

———. "Cosmographers of Seville: Nautical Science and Social Experience." In *First Images of America: The Impact of the New World on the Old,* ed. Fredi Chiappelli, vol. 2, 675–686. Berkeley: University of California Press, 1976.

Langue, Frédérique, and Carmen Salazar Solero. *Dictionnaire des termes miniers en usage en Amérique espagnole (XVIe–XIXe siècle)/Diccionario de términos mineros para la América española (siglos XVI–XIX).* Paris: Editions Recherche sur les Civilisations, 1993.

Latorre, Germán. "Diego Ribero: Cosmógrafo y cartógrafo de la Casa de la Contratación de Sevilla." *Boletín del Centro de Estudios Americanistas* (1918): 27–31.

———. *Diego Ribero: Cosmógrafo y cartógrafo de la Casa de la Contratación de Sevilla.* Seville: Tip. Zarzuela, 1919.

Latour, Bruno. *Science in Action: How to Follow Scientists and Engineers through Society.* Cambridge, Mass: Harvard University Press, 1987.

Law, John. "On the Methods of Long-Distance Control: Vessels, Navigation and the Portuguese Route to India." *Sociological Review Monograph* 32 (1986): 234–263.

León Pinelo, Antonio de, and Andrés González de Barcía. *Epitome de la biblioteca oriental y occidental, náutica y geográfica: Añadido y enmendado nuevamente. . . .* 3 vols. Madrid: Gráficas Yagües, 1973. [Facsimile of the 1737–1738 edition. Madrid: En la Oficina de Francisco Martínez Abad, en la Calle del Olivo Baxo.]

León Portilla, Miguel, ed. *The Broken Spears: The Aztec Account of the Conquest of Mexico.* Boston: Beacon Press, 1992.

Lestringant, Frank. *Mapping the Renaissance World: The Geographical Imagination in the Age of Discovery.* Trans. David Fausett. Berkeley and Los Angeles: University of California Press, 1994.

Lockard, James. *Spanish Peru: 1534–1560: A Social History.* Madison: University of Wisconsin Press, 1994. [1st ed. 1964.]

Lohmann Villena, Guillermo. "La minería en el marco del Virreinato peruano." In *I Coloquio Internacional sobre Historia de la Minería,* vol. 1, 639–655. León, Spain: Cátedra de San Isidoro, 1970.

———. "El Virreinato del Perú." In *El descubrimiento y la fundación de los reinos ultramarinos hasta fines del siglo XVI,* ed. Manuel Lucena Salmoral, vol. 7, 525–562. Madrid: Ediciones Rialp, S.A., 1982.

Long, Pamela O. *Science and Technology in Medieval Society.* Annals 441. New York: New York Academy of Sciences, 1985.

López Piñero, José María. *Ciencia y técnica en la sociedad española de los siglos XVI y XVII.* Barcelona: Editorial Labor, S.A., 1979.

———. *El arte de navegar en la España del Renacimiento.* Barcelona: Editorial Labor, S.A., 1979.

———. "Las 'nuevas medicinas' americanas en la obra (1565–1574) de Nicolás Monardes." *Asclepio* 42, no. 1 (1990): 3–67.

———, ed. *Historia de la ciencia y de la técnica en la Corona de Castilla.* Vol. 3. 4 vols. Salamanca: Junta de Castilla y León Consejería de Educación y Cultura, 2002.

López Piñero, José M., Thomas F. Glick, Víctor Navarro Brotóns, and Eugenio Portela Marco, eds. *Diccionario histórico de la ciencia moderna en España.* Barcelona: Ediciones Península, 1983.

López Piñero, José, and José Pardo Tomás. "The Contribution of Hernández to European Botany and Materia Medica." In *Searching for the Secrets of Nature: The Life and Works of Dr. Francisco Hernández,* ed. Simon Varey et al. Stanford: Stanford University Press, 2000.

Losana Méndez, José. *La sanidad en la época del descubrimiento de América.* Fuenlabrada-Madrid: Cátedra, 1994.

Loscertales, González, Vicente Roldán de Montaud, and Inés Roldán de Montaud. "La minería del cobre en Cuba, su organización, problemas administrativos y repercusiones sociales (1828–1849)." *Revista de Indias* 40 (1980): 255–299.

Lovett, A. W. "Juan de Ovando and the Council of Finance (1573–1575)." *Historical Journal* 15, no. 1 (1972): 1–21.

Lucid, Shannon W. "Six Months on Mir." *Scientific American* 278, no. 5 (May 1998): 46–56.

Luengo Muñoz, Manuel. "Inventos para acrecentar la obtención de perlas en América, durante el siglo XVI." *Anuario de Estudios Americanos* 9 (1952): 51–72.

Lynch, John. *Spain, 1516–1598: From Nation State to World Empire.* Cambridge, Mass.: Blackwell, 1991.

———. *The Hispanic World in Crisis and Change, 1598–1799.* Cambridge, Mass.: Blackwell, 1992.

Maravall, José Antonio. *Estado moderno y mentalidad social (siglos XV a XVII).* Vol. 1. 2 vols. Madrid: Revista de Occidente, 1972.

Medina, José Toribio. *El Veneciano Sebastián Caboto al servicio de España y especialmente de su proyectado viaje a las Molucas por el estrecho de Magallanes y al reconocimiento de la costa del continente hasta la gobernación de Pedrarias Dávila.* 2 vols. Santiago de Chile: Imprenta y Encuadernación Universitaria, 1908.

Melville, Elinor G. K. *A Plague of Sheep.* Cambridge: Cambridge University Press, 1994.

Menes Llaguno, Juan Manuel. *Bartolomé de Medina: Un sevillano pachuquero.* Pachuca, Hidalgo: Universidad Autónoma del Estado de Hidalgo, 1989.

Merchant, Carolyn. *The Death of Nature: Women, Ecology and the Scientific Revolution.* San Francisco: HarperCollins, 1989.

Merton, Robert K. "STS: Foreshadowing of an Evolving Research Program in the Sociology of Science." In *Puritanism and the Rise of Modern Science: The Merton Thesis,* ed. Bernard Cohen. New Brunswick and London: Rutgers University Press, 1990.

Mörner, Magnus. "The Rural Economy and Society of Colonial Spanish South America." In *The Cambridge History of Latin America,* ed. Leslie Bethell, vol. 2. Cambridge: Cambridge University Press, 1997.

Mukerji, Chandra. *From Graven Images: Patterns of Modern Materialism.* New York: Columbia University Press, 1983.

Mundy, Barbary E. *The Mapping of New Spain: Indigenous Cartography and the Maps of the Relaciones Geográficas.* Chicago: University of Chicago Press, 1996.

Muro, Luis. "Bartolomé de Medina, introductor del beneficio de patio en Nueva España." *Historia Mexicana* 13, no. 4 (1964): 517–531.

Nalle, Sara T. *God in La Mancha.* Baltimore and London: Johns Hopkins University Press, 1992.

Nauert, Charles G., Jr. "Caius Plinius Secundus." In *Catalogus Translationum et Commentariorum: Medieval and Renaissance Latin Translations and Commentaries, Annotated Lists and Guides,* ed. Edward Cranz et al., vol. 4. Washington, D.C.: Catholic University of America Press, 1980.

Nummedal, Tara E. "Practical Alchemy and Commercial Exchange in the Holy Roman Empire." In *Merchants and Marvels: Commerce, Science, and Art in Early Modern Europe,* ed. Pamela H. Smith and Paula Findlen. New York: Routledge, 2002.

O'Gorman, Edmundo. *Cuatro historiadores de Indias, siglo XVI: Pedro Mártir de Anglería, Gonzalo Fernández de Oviedo y Valdés, Bartolomé de las Casas, Joseph de Acosta.* Mexico City: Secretaría de Educación Pública, 1972.

Otte, Enrique. "Aspiraciones y actividades heterogéneas de Gonzalo Fernández de Oviedo, cronista." *Revista de Indias* 71 (1958): 9–62.

———. "El proceso del rastro de perlas de Luis de Lampiñan." *Boletín de la Academia Nacional de la Historia de Caracas* 187 (1964): 386–406.

Pagden, Anthony. *The Fall of Natural Man.* Cambridge and New York: Cambridge University Press, 1990.

———. *European Encounters with the New World from the Renaissance to Romanticism.* New Haven: Yale University Press, 1993.

Palau y Dulcet, Antonio. *Manual del librero hispano-americano.* 28 vols. Barcelona: Librería Anticuaria de A. Palau, 1948–1977.

Parker, Geoffrey. "Maps and Ministers: The Spanish Habsburgs." In *Monarchs, Ministers, and Maps: The Emergence of Cartography as a Tool of Government in Early Modern Europe,* ed. David Buisseret. Chicago and London: University of Chicago Press, 1992.

Parry, J. H. *The Discovery of the Sea.* New York: Dial Press, 1974.

———. *The Spanish Seaborne Empire.* Berkeley: University of California Press, 1990.

Paso y Troncoso, Francisco del. *Epistolario de Nueva España: 1505–1818.* 16 vols. Mexico City: Antigua Librería Robredo, de José Porrúa e Hijos, 1940.

Pelorson, Jean Marc. *Les "Letrados": Juristes castillans sous Philippe III: Recherches sur leur place dans la société, la culture et l'état.* Le Puy-en-Velay: l'Eveil de la Haute-Loire, 1980.

Peña Cámara, José María de la, ed. *Contribuciones documentales y críticas para una biografía de Gonzalo Fernández de Oviedo.* Madrid: Revista de Indias, 1957.

Penrose, Boies. *Tudor and Early Stuart Voyaging.* Charlottesville: University Press of Virginia, 1968. [1st ed. 1962.]

Pérez, Juan. "Memoria de las obras y libros de Hernando Colón." In *"Memoria de las obras y libros de Hernando Colón" del bachiller Juan Pérez,* ed. Tomás Marín-Martínez. Madrid: Consejo Superior de Investigaciones Científicas de Sevilla, 1970.

Pérez de Tudela, Juan. "Vida y escritos de Gonzalo Fernández de Oviedo." In *Historia general y natural de las Indias,* vol. 1. Madrid: Biblioteca de Autores Españoles, 1959.

Pérez-Mallaína Bueno, Pablo Emilio. *Los inventos llevados de España a las Indias en la segunda mitad del siglo XVI.* Separata de Cuadernos de Investigación Histórica No. 7. Madrid: Fundación Universitaria Española, 1983.

Phillips, Carla Rahn. *Six Galleons for the King of Spain: Imperial Defense in the Early Sixteenth Century.* Baltimore: Johns Hopkins University Press, 1986.

Picatoste y Rodríguez, Felipe. *Apuntes para una biblioteca científica española del siglo XVI: Estudios biográficos y bibliográficos de ciencias exactas, físicas y naturales y sus inmediatas aplicaciones en dicho siglo.* Madrid: Imprenta de M. Tello, 1891.

Piernas Hurtado, José Manuel. *La Casa de la Contratación de las Indias.* Madrid: Librería de don Victoriano Suárez, 1907.

Pomeranz, Kenneth. *The Great Divide: China, Europe, and the Making of the Modern World Economy.* Princeton and Oxford: Princeton University Press, 2000.

Poole, Stafford. "Juan de Ovando's Reform of the University of Alcalá de Henares, 1564–1566." *Sixteenth Century Journal* 21, no. 4 (1990): 575–606.

Posthumus, N. W. *Inquiry into the History of Prices in Holland.* 2 vols. Leiden: E. J. Brill, 1946.

Pratt, Mary Louise. *Imperial Eyes: Travel Writing and Transculturation.* London and New York: Routledge, 1992.

Puente y Olea, Manuel de la. *Los trabajos geográficos de la Casa de Contratación.* Seville: Escuela Tipográfica y Librería Salesianas, 1900.

Pulido Rubio, José. *El piloto mayor de la Casa de la Contratación de Sevilla: Pilotos mayores, catedráticos de cosmografía y cosmógrafos.* Seville: Escuela de Estudios Hispano-Americanos de Sevilla, 1950.

Ramos Pérez, Demetrio. "Ordenación de la minería en Hispanoamérica durante la época provincial (siglos XVI, XVII, XVIII)." In *I Coloquio Internacional sobre Historia de la Minería,* vol. 1, 373–397. León, Spain: Cátedra de San Isidoro, 1970.

———. "La Junta Magna y la nueva política." In *El descubrimiento y la fundación de los reinos ultramarinos hasta fines del siglo XVI,* ed. Manuel Lucena-Salmoral, vol. 7. Madrid: Ediciones Rialp, S.A., 1982.

Real Academia Española. *Diccionario de autoridades.* Facsimile ed. Madrid: Editorial Gredos, 1963.

Río Moreno, Justo Lucas de. *Los inicios de la agricultura europea en el Nuevo Mundo (1492–1542).* Seville: ASAJA-Sevilla, Caja Rural de Huelva, and Caja Rural de Sevilla, 1991.

Romano, David. *La ciencia hispanojudía.* Madrid: Mapfre, 1992.

Rossi, Paolo. *Philosophy, Technology, and the Arts in the Early Modern Era.* Trans. Salvator Attanasio. New York: Harper and Row, 1970.

Rouse, Irving. *The Tainos: Rise and Decline of the People Who Greeted Columbus.* New Haven: Yale University Press, 1992.

Rumeu de Armas, Antonio. *Hernando Colón, historiador del descubrimiento de América.* Madrid: Instituto de Cultura Hispánica, 1973.

Russell-Wood, A. J. R. *The Portuguese Empire, 1415–1808.* Baltimore: Johns Hopkins University Press, 1998.

Sánchez Cantón, Francisco Javier. *La librería de Juan de Herrera.* Madrid: Instituto Diego Velázquez, 1941.

Sánchez Gómez, Julio. "La técnica en la producción de metales monedables en España y en América, 1500–1650." In *La savia del imperio: Tres estudios de economía colonial,* ed. Julio Sánchez Gómez. Salamanca: Ediciones Universidad de Salamanca, 1997.

Sandman, Alison. "Mirroring the World: Sea Charts, Navigations, and Territorial Claims in Sixteenth-Century Spain." In *Merchants and Marvels: Commerce,*

Science, and Art in Early Modern Europe, ed. Pamela H. Smith and Paula Findlen, 83–108. New York: Routledge, 2002.

Santamaría, Francisco. *Diccionario de mejicanismos.* Mexico City: Ed. Porrúa, 1978.

Sanz López, Carlos, ed. *Relaciones geográficas de España y de Indias.* Madrid: Bibliotheca Americana Vetustissima, 1962.

Sarabia Viejo, María Justina. *Don Luis de Velasco, virrey de Nueva España (1550–1564).* Seville: Escuela de Estudios Hispano-Americanos, 1978.

Sarton, George. "Ptolemy and His Time." In *Ancient Science and Modern Civilization,* 37–73. Lincoln: University of Nebraska Press, 1954.

Schäfer, Ernst. "El origen del Consejo de Indias." *Investigación y Progreso* 7, no. 5 (May 1933): 141–145.

———. "Antonio de Villasante, descubridor droguista en la isla Española." *Investigación y Progreso* 9, no. 1 (1935): 13–15.

———. *El Consejo Real y Supremo de las Indias.* Trans. Ernst Schäfer. Publicaciones del Centro de Estudios de Historia de América. Vol. 1. 2 vols. Seville: Imp. M. Carmona, 1935.

Schmitt, Charles B. "Experience and Experiment: A Comparison of Zabarella's View with Galileo's in *De Motu.*" *Studies in the Renaissance* 16 (1969): 80–138.

Seed, Patricia. *Ceremonies of Possession in Europe's Conquest of the New World, 1492–1640.* Cambridge: Cambridge University Press, 1995.

Sellés, Manuel. *Instrumentos de navegación: Del Mediterráneo al Pacífico.* Barcelona: Lunwerg Editores, 1994.

Serrão, Joel, ed. *Dicionário de história de Portugal.* Lisbon: Iniciativas Editoriais, 1979.

Sevilla Soler, Rosario. "La minería americana y la crisis del siglo XVII: Estado del problema." *Anuario de Estudios Americanos,* special supplement 47, no. 2 (1990): 61–81.

Shapin, Steven. "History of Science and Its Sociological Reconstructions." *History of Science* 20 (1982): 157–211.

———. "The House of Experiment in Seventeenth-Century England." *Isis* 79 (1988): 373–404.

———. *The Scientific Revolution.* Chicago: University of Chicago Press, 1996.

Sigüenza, José de. *Historia de la Orden de San Jerónimo.* 2 vols. Madrid: Bailly-Bailliére e Hijos, 1909.

Smith, Pamela H. *The Business of Alchemy.* Princeton: Princeton University Press, 1994.

Smith, Pamela H., and Paula Findlen, eds. *Merchants and Marvels: Commerce, Science, and Art in Early Modern Europe.* New York: Routledge, 2002.

Solano, Francisco de, ed. *Cuestionarios para la formación de las relaciones geográficas de Indias: Siglos XVI–XIX.* Mi Biblioteca. Madrid: Consejo Superior de Investigaciones Científicas, 1988.

Somolinos D'Ardois, Germán. *Vida y obra de Francisco Hernández.* Mexico City: Universidad Nacional de México, 1960.

———. *El doctor Francisco Hernández y la primera expedición científica en América.* Mexico City: Secretaría de Educación Pública, 1971.

Stern, Steve J. *Peru's Indian Peoples and the Challenge of Spanish Conquest.* Madison: University of Wisconsin Press, 1993. [1st ed. 1982.]

Stradling, R. A. "Seventeenth Century Spain: Decline or Survival?" *European Studies Review* 9 (1979): 157–194.

Tachot, Louise Bénet. "Comentarios acerca de la trayectoria editorial de *La Historia General y Natural de las Indias* de Gónzalo Fernández de Oviedo y Valdés." Unpublished paper, 49th International Congress of Americanists, Quito, Ecuador, 1997.

Thompson, I. A. A. *War and Society in Habsburg Spain: Select Essays.* Aldershot, Hampshire: Variorum, 1992.

Turner, Gerard L'E. *Scientific Instruments, 1500–1900: An Introduction.* Berkeley, Los Angeles, and London: Philip Wilson and University of California Press, 1998.

Valverde, José Luis, Teresa Bautista, and María Teresa Montaña. *Libros de interés histórico-médico-farmacéutico conservados en la Biblioteca de la Real Academia de Medicina de Sevilla.* Granada: Universidad de Granada, 1980.

Varey, Simon, ed. *The Mexican Treasury: The Writings of Dr. Francisco Hernández.* Stanford: Stanford University Press, 2000.

Varey, Simon, Rafael Chabrán, and Dora B. Weiner, eds. *Searching for the Secrets of Nature: The Life and Works of Dr. Francisco Hernández.* Stanford: Stanford University Press, 2000.

Vas Mingo, Milagros de. *Las capitulaciones de Indias en el siglo XVI.* Madrid: Instituto de Cooperación Iberoamericana, 1986.

Vassberg, David. *Land and Society in Golden Age Castile.* New York: Cambridge University Press, 1984.

Vernet Gines, Juan. *Historia de la ciencia española.* Madrid: Instituto de España: Cátedra "Alfonso X el Sabio," 1975.

Vos, Paula De. "Spain in the 'New World' of Europe: The Casa de la Contratación and the Organization of Navigation and Cartography in the Sixteenth Century." Master's thesis. University of California, Berkeley, 1993.

Wagner, Klaus. *Catálogo abreviado de las obras impresas del siglo XVI de la Biblioteca Universitaria de Sevilla.* Seville: Universidad de Sevilla, 1988.

Waters, D. W. *The Art of Navigation.* New Haven: Yale University Press, 1958.

Weber, Max. *The Protestant Ethic and the Spirit of Capitalism.* Trans. Talcott Parsons. New York: Charles Scribner's Sons, 1958.

White, Lynn. *Medieval Technology and Social Change.* Oxford: Oxford University Press, 1962.

Wilkinson-Zerner, Catherine. *Juan de Herrera: Architect to Philip II of Spain.* New Haven: Yale University Press, 1993.

Zilsel, Edgar. "The Sociological Roots of Science." In *The Social Origins of Modern Science,* ed. Diederick Raven et al. Dordrecht, Boston, and London: Kluwer Academic Publishers, 2000.

Index

Printed and bound by CPI Group (UK) Ltd, Croydon, CR0 4YY

13/04/2025

14656492-0002